PE Mechanical

Thermal and Fluids Systems Practice Exam

Michael R. Lindeburg, PE

PPI®
PPI2PASS.COM
A KAPLAN COMPANY

PE MECHANICAL THERMAL AND FLUIDS SYSTEMS PRACTICE EXAM

Current release of this edition: 2

Release History

date	edition number	revision number	update
Oct 2017	1	1	New book.
Jan 2019	1	2	Minor corrections.

Printed in the United States of America.

PPI
1250 Fifth Avenue, Belmont, CA 94002
(650) 593-9119
ppi2pass.com

ISBN: 978-1-59126-542-9

Library of Congress Control Number: 2017954340

F E D C B A

Table of Contents

Preface and Acknowledgments

By this time, if you are at a stage in your preparation where you are picking up *PE Mechanical Thermal and Fluids Systems Practice Exam* and reading this preface (most people don't), then you probably have already read the prefaces in the *Mechanical Engineering Reference Manual* and *Practice Problems for the Mechanical Engineering PE Exam.* In that case, you will have read all the clever and witty things I wrote in those prefaces, so I won't be able to reuse them here.

This book accurately reflects the NCEES Mechanical—Thermal and Fluids Systems exam specifications and multiple-choice format. The format of the PE exam presents several challenges to examinees. The breadth of the subject matter covered by the exam requires that you have a firm grasp of mechanical engineering fundamentals. These rudiments are generally covered in an undergraduate curriculum. You still need to know how to select a pump, size a shaft, and humidify a room. However, exam problems enable NCEES to focus on, target, and drill down to some very specific knowledge bases. The problems within each area of emphasis require knowledge gained only through experience—a true test of your worthiness of licensure. Often, problems testing proficiency in an area can be worded as simple definition questions. For example, if you don't recognize the application of a jockey pump, you probably aren't ready to design fire protection sprinkler systems.

Needless to say, the number of problems in this practice exam is consistent with the NCEES exam, with 80 total problems. Like the exam, this book exclusively uses U.S. customary system (USCS) units. The solutions in this practice exam are also consistent in nomenclature and style with the *Mechanical Engineering Reference Manual.* Hopefully, you have already made that book part of your exam-day arsenal.

While *PE Mechanical Thermal and Fluids Systems Practice Exam* reflects the NCEES exam specifications, none of the problems in this book are actual exam problems. The problems in this book have come directly from my imagination and the imagination of my colleagues, Keith Elder, PE, and James Kamm, PhD, PE, LEED AP. Material was technically reviewed by Anthony Darmiento, PE. I am grateful for their efforts.

Editing, typesetting, illustrating, and proofreading of this book continues to follow PPI's strict style guidelines for engineering publications. Managed by Grace Wong, director of publishing services, the following top-drawer individuals have my thanks for bringing this book to life: Nancy Peterson, project manager; Cathy Schrott, production services manager; Jenny Lindeburg King, editor-in-chief; Scott Marley, copy editor; Scott Rutherford, proofreader; Darren Meiss, freelance proofreader; Robert Genevro and Ellen Nordman, typesetters; Tom Bergstrom, technical illustrator and cover designer; and Anil Acharya, calculation checker. Without their efforts, this book would read like a bunch of scribbles.

As in all of my publications, I invite your comments. If you disagree with a solution, or if you think there is a better way to do something, please let me know. You can submit errata online at the PPI website at **ppi2pass.com/errata**.

Best wishes in your exam and subsequent career.

Michael R. Lindeburg, PE

Codes Used to Prepare This Book

The documents, codes, and standards that I used to prepare this book were the most current available at the time of publication. In the absence of any other specific need, that was the best strategy.

Engineering practice is often constrained by law or contract to using codes and standards that have already been adopted or approved. However, newer codes and standards might be available. For example, the adoption of building codes by states and municipalities often lags publication of those codes by several years. By the time the 2015 codes are adopted, the 2017 codes have been released. Federal regulations are always published with future implementation dates. Contracts are signed with designs and specifications that were "best practice" at some time in the past. Nevertheless, the standards are referenced by edition, revision, or date. All of the work is governed by unambiguous standards.

All standards produced by ASME, ASHRAE, ANSI, ASTM, and similar organizations are identified by an edition, revision, or date. However, although NCEES lists "codes and standards" in its lists of Mechanical—Thermal and Fluids Systems exam specifications, no editions, revisions, or dates are specified. My conclusion is that the exam is not sensitive to changes in codes, standards, regulations, or announcements in the Federal Register. This is the reason I referred to the most current documents available as I prepared this book.

Introduction

ABOUT THE PE EXAM

PE Mechanical Thermal and Fluids Systems Practice Exam provides the opportunity to practice taking an eight-hour exam similar in content and format to the Principles and Practice of Engineering (PE) Mechanical—Thermal and Fluids Systems exam. The exam is eight-hours divided into a morning session and an afternoon session. In the four-hour morning session, you are asked to solve 40 problems and in the four-hour afternoon session you are also asked to solve 40 problems. All in all, you will be solving 80 problems in eight hours.

All problems are multiple-choice and in U.S. customary system (USCS) units. They include a problem statement with all the required information, followed by four logical choices. Only one of the four options is correct. The problems are completely independent from each other, so an incorrect choice on one problem will not carry over to subsequent problems.

The problems for each session are drawn from either the Principles or Applications knowledge areas as specified by the National Council of Examiners for Engineering and Surveying (NCEES). The exam knowledge areas and the approximate number of problems are as follows.

Principles (32 problems)

- Basic Engineering Practice (6 problems)
- Fluid Mechanics (6 problems)
- Heat Transfer Principles (6 problems)
- Mass Balance Principles (4 problems)
- Thermodynamics (6 problems)
- Supportive Knowledge (4 problems)

Hydraulic and Fluid Applications (24 problems)

- Hydraulic and Fluid Equipment (15 problems)
- Distribution Systems (9 problems)

Energy/Power System Applications (24 problems)

- Energy/Power Equipment (8 problems)
- Cooling/Heating (6 problems)
- Energy Recovery (5 problems)
- Combined Cycles (5 problems)

HOW TO USE THIS BOOK

This book is a practice exam—the main issue is not how you use it, but when you use it. It was not intended to be a diagnostic tool to guide your preparation. Rather, its value is in giving you an opportunity to bring together all of your knowledge and to practice your test-taking skills. The three most important skills are (1) selection of the right subjects to study, (2) organization of your references and other resources, and (3) time management. Take this practice exam within a few weeks of your actual exam. That's the only time that you will be able to focus on test-taking skills without the distraction of rusty recall.

Do not read the questions ahead of time, and do not look at the answers until you've finished. Prepare for the practice exam as you would prepare for the actual exam. Assemble your reference materials. Check with your state's board of engineering registration for any restrictions on what materials you can bring to the exam. (The PPI website has a listing of state boards at **ppi2pass.com/faqs/state-boards**.)

Read the practice exam instructions (which simulate the ones you'll receive from your exam proctor), set a timer for four hours, and answer the first 40 problems. After a one-hour break, turn to the last 40 problems in this book, set the timer, and complete the simulated afternoon session. Then, check your answers.

The problems in this book were written to emphasize the breadth of the thermal and fluids systems mechanical engineering field. Some may seem easy and some hard. If you are unable to answer a problem, you should review that topic area in the *Mechanical Engineering Reference Manual for the PE Exam.*

The problems are generally similar to each other in difficulty, yet a few somewhat easier problems have been included to expose you to less frequently examined topics. On the exam, only your answer sheet will be scored. No credit will be given for calculations written in your exam booklet.

While some topics in this book may not appear on your exam, the concepts and problem style will be useful practice.

Once you've taken the full exam, check your answers. Evaluate your strengths and weaknesses, and select additional texts to supplement your weak areas (e.g., *Practice Problems for the Mechanical Engineering PE Exam*). Check the PPI website for the latest in exam preparation materials at **ppi2pass.com**.

The keys to success on the exam are to know the basics and to practice solving as many problems as possible. This book will assist you with both objectives.

Pre-Test

You can use the following incomplete table to judge your preparedness. You should be able to fill in all of the missing information. (The completed table appears on the back of this page.) If you are ready for the sample exam (and, hence, for the actual exam), you will recognize them all, get most correct, and when you see the answers to the ones you missed, you'll say, "Ahh, yes." If you have to scratch your head with too many of these, then you haven't exposed yourself to enough of the subjects that are on the exam.

description	value or formula	units
acceleration of gravity, g		in/sec^2
gravitational constant, g_c	32.2	
formula for the area of a circle		ft^2
	1545	ft-lbf/lbmol-°R
	+460	°
density of water, approximate		lbm/ft^3
density of air, approximate	0.075	
specific gas constant for air	53.35	
	1.0	Btu/lbm-°R
foot-pounds per second in a horsepower		ft-lbf/hp-sec
pressure, p, in fluid with density, ρ, in lbm/ft^3, at depth, h		lbf/ft^2
cancellation and simplification of the units (A = amps; rad = radians)	A-sec^4/sec^5-rad	
specific heat of air, constant pressure		Btu/lbm-°R
common units of entropy of steam	–	
	$bh^3/12$	cm^4
molecular weight of oxygen gas		lbm/lbmol
what you add to convert $\Delta T_{°F}$ to $\Delta T_{°R}$		°
	Q/A	ft/sec
inside surface area of a hollow cylinder with length L and diameters d_i and d_o		ft^2
the value that NPSHA must be larger than		ft
	849	lbm/ft^3
what you have to multiply density in lbm/ft^3 by to get specific weight in lbf/ft^3		lbf/lbm
power dissipated by a device drawing I amps when connected to a battery of V volts		W
linear coefficient of thermal expansion for steel	6.5×10^{-6}	
formula converting degrees centigrade to degrees Celsius		
the primary SI units constituting a newton of force		N
the difference between psig and psia at sea level		psi
the volume of a mole of an ideal gas		ft^3
	1.71×10^{-9}	Btu/ft^2-hr-°R^4
universal gas constant	8314.47	
	2.31	ft/psi
shear modulus of steel		psi
conversion from rpm to rad/sec		rad-min/rev-sec
Joule's constant	778.17	

Pre-Test Answer Key

description	value or formula	units
acceleration of gravity, g	386	in/sec^2
gravitational constant, g_c	32.2	lbm-ft/lbf-sec^2
formula for the area of a circle	πr^2 or $(\pi/4)d^2$	ft^2
universal gas constant in customary U.S. units	1545	ft-lbf/lbmol-°R
what you add to $T_{°F}$ to obtain $T_{°R}$ (absolute temperature)	+460	°
density of water, approximate	62.4	lbm/ft^3
density of air, approximate	0.075	lbm/ft^3
specific gas constant for air	53.35	ft-lbf/lbm-°R
specific heat of water	1.0	Btu/lbm-°R
foot-pounds per second in a horsepower	550	ft-lbf/hp-sec
pressure, p, in fluid with density, ρ, in lbm/ft^3, at depth, h	$p=\gamma h=\rho gh/g_c$	lbf/ft^2
cancellation and simplification of the units (A = amps; rad = radians)	A-sec^4/sec^5-rad	W (watts)
specific heat of air, constant pressure	0.241	Btu/lbm-°R
common units of entropy of steam	–	Btu/lbm-°R
centroidal moment of inertia of a rectangle	$bh^3/12$	cm^4
molecular weight of oxygen gas	32	lbm/lbmol
what you add to convert $\Delta T_{°F}$ to $\Delta T_{°R}$	0	°
velocity of flow	Q/A	ft/sec
inside surface area of a hollow cylinder with length L and diameters d_i and d_o	$\pi d_i L$	ft^2
the value that NPSHA must be larger than	h_v (vapor head)	ft
density of mercury	849	lbm/ft^3
what you have to multiply density in lbm/ft^3 by to get specific weight in lbf/ft^3	g/g_c (numerically, 32.2/32.2 or 1.0)	lbf/lbm
power dissipated by a device drawing I amps when connected to a battery of V volts	IV	W
linear coefficient of thermal expansion for steel	6.5×10^{-6}	1/°F or 1/°R
formula converting degrees centigrade to degrees Celsius	°centigrade = °Celsius	The centigrade scale is obsolete.
the primary SI units constituting a newton of force	kg·m/s^2	N
the difference between psig and psia at sea level	14.7 psia (atmospheric pressure)	psi
the volume of a mole of an ideal gas	359 or 360	ft^3
Stefan-Boltzmann constant	1.71×10^{-9}	Btu/ft^2-hr-°R^4
universal gas constant	8314.47	J/kmol·K
conversion from psi to height of water	2.31	ft/psi
shear modulus of steel	11.5×10^6	psi
conversion from rpm to rad/sec	$2\pi/60$	rad-min/rev-sec
Joule's constant	778.17	ft-lbf/Btu

Practice Exam Instructions

In accordance with the rules established by your state, you may use textbooks, handbooks, bound reference materials, and any approved battery- or solar-powered, silent calculator to work this examination. However, no blank papers, writing tablets, unbound scratch paper, or loose notes are permitted. Sufficient room for scratch work is provided in the Examination Booklet.

You are not permitted to share or exchange materials with other examinees.

You will have eight hours for the examination: four hours to answer the first 40 questions, a one hour lunch break, and four hours to answer the second 40 questions. Your score will be determined by the number of questions that you answer correctly. There is a total of 80 questions. All 80 questions must be worked correctly in order to receive full credit on the exam. There are no optional questions. Each question is worth 1 point. The maximum possible score for the examination is 80 points.

Partial credit is not available. No credit will be given for methodology, assumptions, or work written in your Examination Booklet.

Record all of your answers on the Answer Sheet. No credit will be given for answers marked in the Examination Booklet. Mark your answers with a no. 2 pencil. Answers marked in pen may not be graded correctly. Marks must be dark and must completely fill the bubbles. Record only one answer per question. If you mark more than one answer, you will not receive credit for the question. If you change an answer, be sure the old bubble is erased completely; incomplete erasures may be misinterpreted as answers.

If you finish early, check your work and make sure that you have followed all instructions. After checking your answers, you may turn in your Examination Booklet and Answer Sheet and leave the examination room. Once you leave, you will not be permitted to return to work or change your answers.

When permission has been given by your proctor, break the seal on the Examination Booklet. Check that all pages are present and legible. If any part of your Examination Booklet is missing, your proctor will issue you a new Booklet.

WAIT FOR PERMISSION TO BEGIN.

Name: ______________________________________
Last First Middle Initial

Examinee number: ______________

Examination Booklet number: ______________

Principles and Practice of Engineering Examination

Practice Examination

1.	(A)	(B)	(C)	(D)
2.	(A)	(B)	(C)	(D)
3.	(A)	(B)	(C)	(D)
4.	(A)	(B)	(C)	(D)
5.	(A)	(B)	(C)	(D)
6.	(A)	(B)	(C)	(D)
7.	(A)	(B)	(C)	(D)
8.	(A)	(B)	(C)	(D)
9.	(A)	(B)	(C)	(D)
10.	(A)	(B)	(C)	(D)
11.	(A)	(B)	(C)	(D)
12.	(A)	(B)	(C)	(D)
13.	(A)	(B)	(C)	(D)
14.	(A)	(B)	(C)	(D)
15.	(A)	(B)	(C)	(D)
16.	(A)	(B)	(C)	(D)
17.	(A)	(B)	(C)	(D)
18.	(A)	(B)	(C)	(D)
19.	(A)	(B)	(C)	(D)
20.	(A)	(B)	(C)	(D)
21.	(A)	(B)	(C)	(D)
22.	(A)	(B)	(C)	(D)
23.	(A)	(B)	(C)	(D)
24.	(A)	(B)	(C)	(D)
25.	(A)	(B)	(C)	(D)
26.	(A)	(B)	(C)	(D)
27.	(A)	(B)	(C)	(D)
28.	(A)	(B)	(C)	(D)
29.	(A)	(B)	(C)	(D)
30.	(A)	(B)	(C)	(D)
31.	(A)	(B)	(C)	(D)
32.	(A)	(B)	(C)	(D)
33.	(A)	(B)	(C)	(D)
34.	(A)	(B)	(C)	(D)
35.	(A)	(B)	(C)	(D)
36.	(A)	(B)	(C)	(D)
37.	(A)	(B)	(C)	(D)
38.	(A)	(B)	(C)	(D)
39.	(A)	(B)	(C)	(D)
40.	(A)	(B)	(C)	(D)
41.	(A)	(B)	(C)	(D)
42.	(A)	(B)	(C)	(D)
43.	(A)	(B)	(C)	(D)
44.	(A)	(B)	(C)	(D)
45.	(A)	(B)	(C)	(D)
46.	(A)	(B)	(C)	(D)
47.	(A)	(B)	(C)	(D)
48.	(A)	(B)	(C)	(D)
49.	(A)	(B)	(C)	(D)
50.	(A)	(B)	(C)	(D)
51.	(A)	(B)	(C)	(D)
52.	(A)	(B)	(C)	(D)
53.	(A)	(B)	(C)	(D)
54.	(A)	(B)	(C)	(D)
55.	(A)	(B)	(C)	(D)
56.	(A)	(B)	(C)	(D)
57.	(A)	(B)	(C)	(D)
58.	(A)	(B)	(C)	(D)
59.	(A)	(B)	(C)	(D)
60.	(A)	(B)	(C)	(D)
61.	(A)	(B)	(C)	(D)
62.	(A)	(B)	(C)	(D)
63.	(A)	(B)	(C)	(D)
64.	(A)	(B)	(C)	(D)
65.	(A)	(B)	(C)	(D)
66.	(A)	(B)	(C)	(D)
67.	(A)	(B)	(C)	(D)
68.	(A)	(B)	(C)	(D)
69.	(A)	(B)	(C)	(D)
70.	(A)	(B)	(C)	(D)
71.	(A)	(B)	(C)	(D)
72.	(A)	(B)	(C)	(D)
73.	(A)	(B)	(C)	(D)
74.	(A)	(B)	(C)	(D)
75.	(A)	(B)	(C)	(D)
76.	(A)	(B)	(C)	(D)
77.	(A)	(B)	(C)	(D)
78.	(A)	(B)	(C)	(D)
79.	(A)	(B)	(C)	(D)
80.	(A)	(B)	(C)	(D)

Practice Exam

1. Based on chemical resistance alone, which of the following pipe materials would provide the best performance?

(A) polyvinyl chloride (PVC)

(B) polystyrene

(C) polypropylene

(D) acrylonitrile butadiene styrene (ABS)

2. The water level of a 15 psig pressurized tank is 8 ft below the water level of an open tank. A pump 15 ft below the water level of the open tank delivers water through schedule-40 steel pipe to the pressurized tank. The losses due to both pipe friction and fittings are 12 ft. The pump must have a total dynamic head that is most nearly

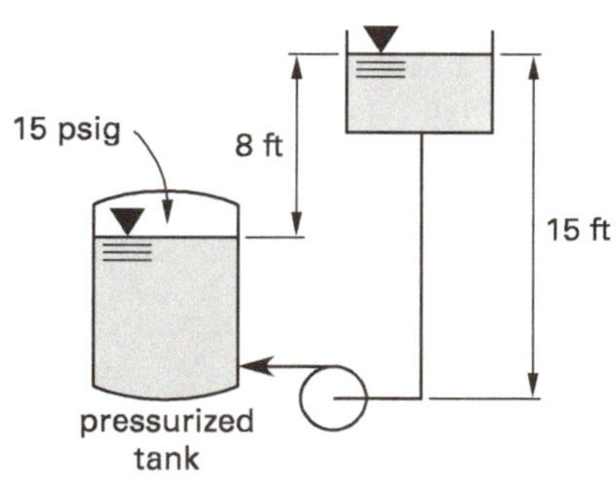

(A) 19 ft water

(B) 27 ft water

(C) 39 ft water

(D) 54 ft water

3. When running at full load, a chiller has the capacity to cool 640 gpm of water from 55°F to 43°F. If the rated coefficient of performance (COP) is 4.4, the total compressor heat that must be rejected to the cooling tower is most nearly

(A) 21 tons

(B) 73 tons

(C) 320 tons

(D) 390 tons

4. The illustration shown is an example of a(n)

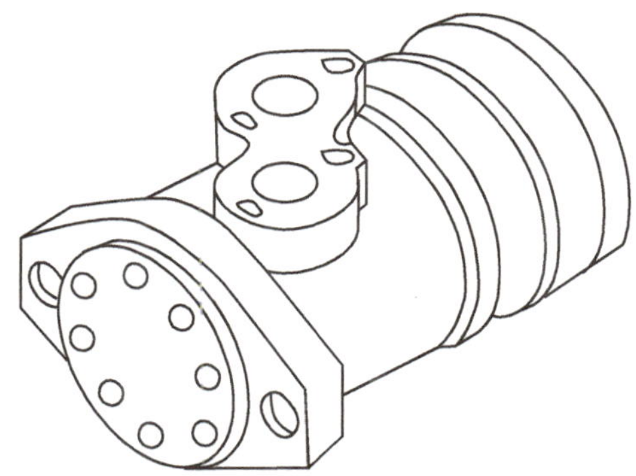

(A) cavalier view

(B) isometric view

(C) principal view

(D) sectional view

5. A jet of compressed air is used to bring a large gen-set (a turbine-powered emergency electrical generator) up to speed in 3 sec. The compressed air acts perpendicularly to the turbine's longitudinal axis of rotation. The effective torque moment arm (causing rotation) of the jet is 1.5 ft. The four-pole generator is direct-driven at 188.5 rad/sec. The compressed air is introduced at three times the tangential velocity of the turbine, but the air supply valve closes when the turbine has reached 110% of its steady-state rotational velocity. What is most nearly the initial Mach number of the compressed air used to start the gen-set?

(A) 0.41

(B) 0.54

(C) 0.62

(D) 0.75

6. A hydraulic reservoir with a diameter of 30 in is open to the atmosphere and has a $\frac{1}{8}$ in diameter relief check valve feeding a return line. The flow coefficient for the check valve is 0.53, and the valve's cracking pressure is 5 psi. The reservoir contains 24 in of hydraulic oil having a specific gravity of 0.90 at an operating temperature of 100°F. When the oil pressure drop across the valve is 5.775 psig, what is the approximate flow rate of the hydraulic oil into the reservoir from the return line?

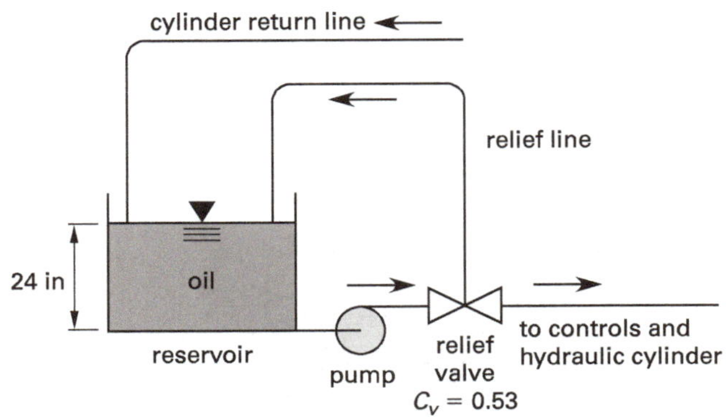

(A) 0.41 gpm

(B) 0.53 gpm

(C) 1.3 gpm

(D) 1.4 gpm

7. The kinematic viscosity of an SAE 10W-30 engine oil with a specific gravity of 0.88 is reported as 110 centistokes (cSt) at 37°C. Expressed in Saybolt universal seconds (SUS), the viscosity is most nearly

(A) 0.90 SUS

(B) 49 SUS

(C) 510 SUS

(D) 32,000 SUS

8. Dry air flows into a long, insulated channel that contains a pool of water. Liquid water flows in at point 2 at the same rate as the evaporation rate to maintain a constant quantity of liquid. The air flows in through the inlet at point 1 at a rate of 12 kg/s at a temperature of 25°C and pressure of 100 kPa. If the air exits at point 3 saturated at 40°C, what is most nearly the channel water's rate of evaporation?

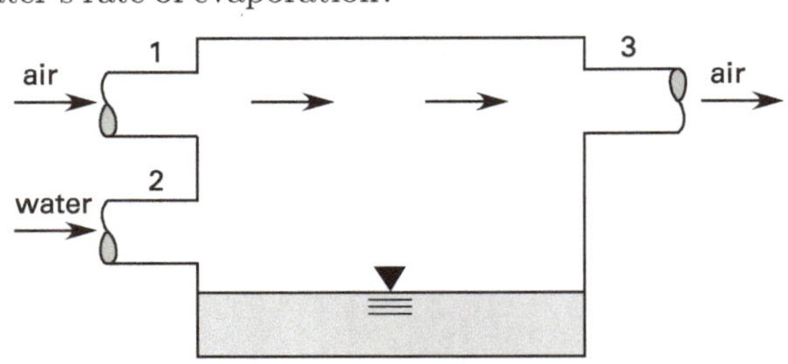

(A) 0.04 kg/s

(B) 0.6 kg/s

(C) 10 kg/s

(D) 100 kg/s

9. A 20 ft length of $2\frac{1}{2}$ in nominal diameter schedule-40 steel pipe full of water is simply supported. What is most nearly the maximum pipe deflection?

(A) 0.18 in

(B) 0.45 in

(C) 0.50 in

(D) 0.61 in

10. A fluid flows through a tube at a rate of 3×10^{-5} ft³/sec. The tube is 3.4 ft long and has an internal diameter of 0.018 in. The flow is viscous with a viscosity of 26.37×10^{-6} lbf-sec/ft², incompressible, steady, and laminar. The predicted pressure drop across the length of the tube is most nearly

(A) 11,000 lbf/ft²

(B) 15,000 lbf/ft²

(C) 22,000 lbf/ft²

(D) 26,000 lbf/ft²

11. A cylindrical tank is produced from alternating layers of a low-ϵ glass, fiber-reinforced epoxy. The fiber orientation angles in alternating layers of the tank wall are at 55° with respect to the principal axes along both the longitudinal and circumferential directions. The tank contains nitrogen gas at a pressure of 10 MPa. The tank has a mean diameter of 300 mm. There is no torsion on the tank. To limit the normal stress to 115 MPa in the direction of fiber orientation, what is most nearly the minimum wall thickness?

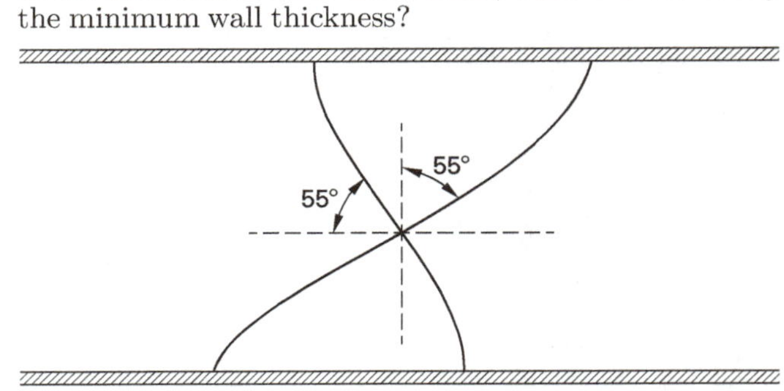

(A) 8.7 mm

(B) 11 mm

(C) 14 mm

(D) 21 mm

12. Investors are evaluating an office building that has a purchase price of $525,000. The investors expect to sell the building for $700,000 after 7 years. During the time that the building is held, the building will be depreciated using straight line depreciation, an estimated life of 15 years, and an (assumed) salvage value of zero.

(The full purchase price will be depreciated.) The effective tax rate of the investors is 40%.

During ownership, the estimated expenses will be \$25,000 per year, which will be offset by rental income of \$45,000 for the first 3 years and \$75,000 for the next 4 years.

Neglect the benefit of a favorable capital gains tax rate on the gain above \$525,000. The after-tax rate of return on this investment will be most nearly

(A) 2.1%

(B) 7.1%

(C) 7.9%

(D) 12%

13. A 20 in diameter suction line carries water at 7000 gpm. The water enters a pump at 100 psi and is pumped up a 30 ft high incline through a 12 in diameter discharge line. If the pressure at the top of the incline is 500 psi and the head loss through the 30 ft section is 10 ft, the horsepower that is supplied by the pump is most nearly

(A) 140 hp

(B) 750 hp

(C) 1200 hp

(D) 1700 hp

14. A 150 lbm cartop carrier measures 51 in long × 35 in wide × 18 in high. If a 3000 lbm car is driven at 65 mph into a 10 mph headwind, the added net motor power required when the carrier is used is most nearly

(A) 5.9 hp

(B) 13 hp

(C) 15 hp

(D) 36 hp

15. Air with a mass of 3 lbm is expanded polytropically behind a piston in a cylinder from an initial condition of 120 psia and 100°F to a final condition of 40 psia. Using a polytropic exponent of 1.6, the work done by the closed system per unit mass is most nearly

(A) -1.7×10^4 ft-lbf/lbm

(B) -5.6×10^3 ft-lbf/lbm

(C) 5.6×10^3 ft-lbf/lbm

(D) 1.7×10^4 ft-lbf/lbm

16. A heat exchanger pump receives water at 140°F and returns it at 95°F. The water is received and returned through nominal 3 in copper tubing (with an outside diameter of 3.5 in, inside diameter of 3.062 in, cross-sectional area of 7.370 in^2, and electrical resistivity of 10 Ω-cmil/ft). The pump is driven by a one-phase, 110 V AC, 5 hp motor that draws an average (RMS) current of 35 A. The engineer decides to place the copper tubing inside PVC pipe for electrical insulation, and use the copper tubing to carry the electrical current that will power the pump motor. The maximum distance that the pump and motor can be located from a 110 V source so that the voltage drop does not exceed 0.6 V is most nearly

(A) 870 ft

(B) 2500 ft

(C) 4900 ft

(D) 9000 ft

17. In the system shown, the flow rate of 68°F water is 1500 gpm, and the head loss (in feet) in the pipes is given by $0.02(L/D)(\text{v}^2/2g)$, where L is the length of pipe in feet, D is the pipe diameter in feet, v is the velocity in ft/sec, and g is the acceleration of gravity in ft/sec^2. If the motor driving the pump is 90% efficient and the pump is 80% efficient, the power that must be supplied to the motor is most nearly

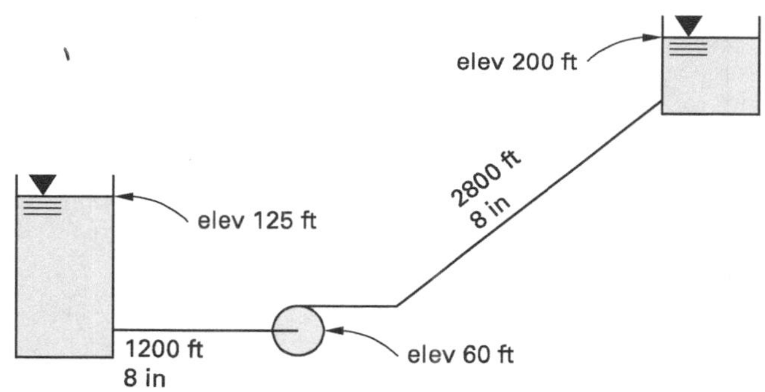

(not to scale)

(A) 52 kW

(B) 69 kW

(C) 97 kW

(D) 130 kW

18. A 2 hp motor is used to stir a tank containing 60 lbm of water for 15 min. Assuming the process occurs at constant volume, the maximum possible rise in temperature is most nearly

(A) 1.4°F

(B) 21°F

(C) 58°F

(D) 130°F

19. What type of fit is indicated by the manufacturing illustration shown?

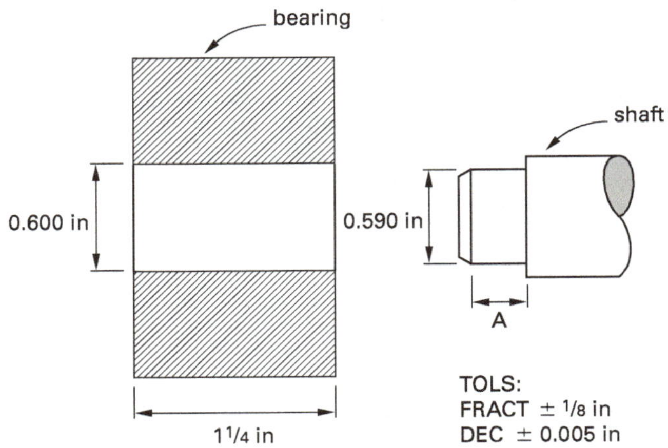

(A) clearance fit

(B) interference fit

(C) press fit

(D) tolerance fit

20. The rectangular wall of a furnace is made from 3 in fire-clay brick ($k = 0.58$ Btu-ft/hr-ft^2-°F) surrounded by 0.25 in of steel on the outside. 0.25 in diameter mild steel bolts connect the steel to the brick. The furnace is surrounded by 70°F air with a convection coefficient of 1.65 Btu/hr-ft^2-°F, while the inner surface of the brick is held constant at 1000°F. Disregard conductance through the bolts. The outside surface temperature of the steel is most nearly

(A) 160°F

(B) 470°F

(C) 610°F

(D) 760°F

21. A parallel-flow tube-and-shell heat exchanger is designed using 1 in OD tubing. 40,000 lbm/hr of water at 55°F are used to cool 45,000 lbm/hr of a 95% ethyl alcohol solution ($c_p = 0.9$ Btu/lbm-°F) from 160°F to 110°F. If the overall coefficient of heat transfer based on the outer tube area is 75 Btu/hr-ft^2-°F, then the heat-transfer surface area of the heat exchanger is most nearly

(A) 520 ft^2

(B) 850 ft^2

(C) 920 ft^2

(D) 1100 ft^2

22. A centrifugal pump with the pump curve given is to be installed as shown. The line uses 16 in schedule-40 steel pipe and contains two long-radius 90° elbows and one gate valve. The pipe friction factor is 0.015. The discharge will be most nearly

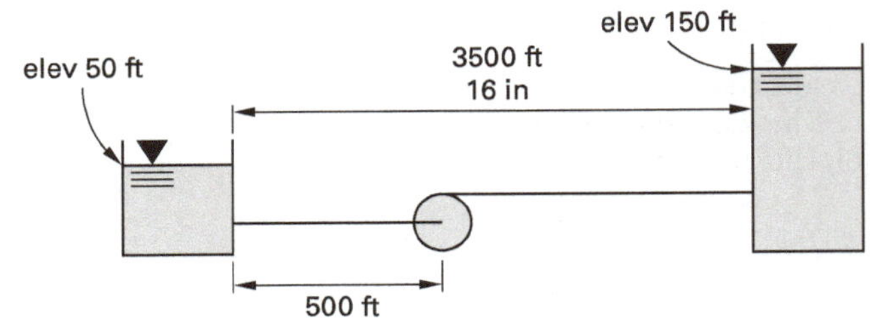

(not to scale)

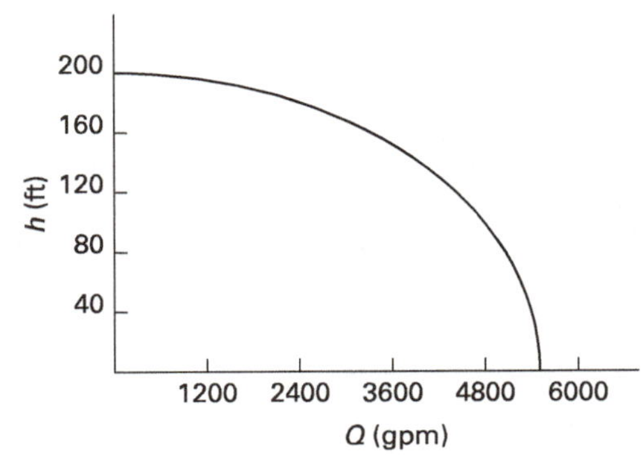

(A) 2400 gpm

(B) 3000 gpm

(C) 4000 gpm

(D) 4200 gpm

23. Water at 20°C flows through the pipe system (19 mm ID) shown at a rate of 185 L/min (0.185 m^3/min). If the loss coefficients for the 90° elbows and 45° elbow are 1.5 and 0.4, respectively, what is most nearly the sum of the minor (head) losses for the system?

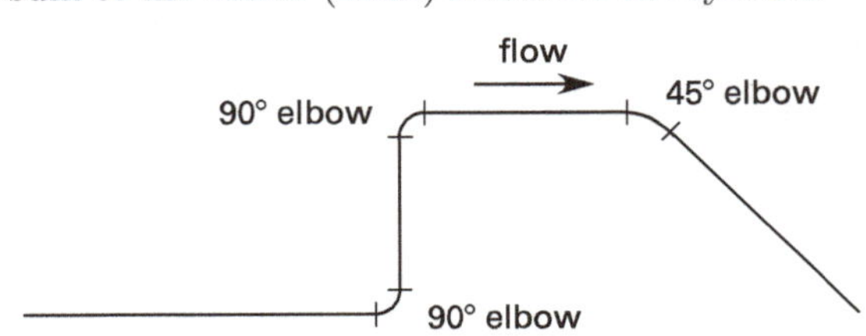

plan view

(A) 2.4 m

(B) 18 m

(C) 21 m

(D) 23 m

24. A steam turbine operates as a component of a Rankine cycle. Steam is supplied to the turbine at 1000 psia and 800°F. The turbine exhausts at 4 psia. The expansion is not reversible, and the exhaust is vapor at 100% quality. The thermal efficiency of the turbine is most nearly

(A) 59%

(B) 76%

(C) 85%

(D) 100%

25. A hydraulic press is used to compress soap cakes into shaped bars. The hydraulic fluid has an operating pressure of 900 psig and a flow rate of 3 gpm, and the press has an 85% efficient cylinder that travels at 0.25 ft/sec. The work done in compressing a tray of soap bars is 50 ft-lbf. Approximately how much is the thickness of each soap cake reduced in compression?

(A) 0.12 in

(B) 0.14 in

(C) 0.17 in

(D) 0.20 in

26. A solid copper sphere 200 mm in diameter is at a temperature of 450°C. It is dropped into a large tank of 75°C oil. If the average convective heat transfer coefficient is 880 W/m²·K and the oil is stirred uniformly at all times, most nearly how long after submersion does the sphere reach a temperature of 200°C?

(A) 5.3 s

(B) 160 s

(C) 910 s

(D) 4700 s

27. An irrigation system is being designed in which a water tower will supply the irrigation ditches. The water is to be discharged at 60 gpm. The gate valve that will control the water discharge is to operate fully open, and the pressure drop across this valve is to be no more than 5 psi. Of the following options, the smallest suitable size for the gate valve is

(A) $1/2$ in

(B) $3/4$ in

(C) 1 in

(D) $1\,1/4$ in

28. Thirty-two balls are contained in a 2 ft wide, 2 ft deep, 3 ft tall container for a lottery drawing. A fan blowing 70°F air, 14.7 psia upward from beneath the container forces the balls upward in a clear tube so as to display the number on each ball. If the balls each have a diameter of 1.5 in and weigh 0.00516 lbf, the flow rate through the container must be at least

(A) 4500 ft³/min

(B) 7200 ft³/min

(C) 11,000 ft³/min

(D) 22,000 ft³/min

29. The power output from a generator that is driven by an impulse turbine is 122 000 kW. The turbine wheel, which is mounted on a 3 ft diameter steel shaft, has a 20 ft diameter and is spun by a 2 ft diameter water jet. Assuming ideal conditions ($v_{jet} = 2v_{vane}$ and the jet is deflected through 180°), the angular speed of the turbine wheel is most nearly

(A) 15 rpm

(B) 150 rpm

(C) 1300 rpm

(D) 2600 rpm

30. In the system shown, the head loss (in feet) in the pipes is given by $0.015(L/D)(v^2/2g)$, where L is the length of pipe in feet, D is the pipe diameter in feet, v is the velocity in ft/sec, and g is the acceleration of gravity in ft/sec². The maximum and minimum pressures in the system are most nearly

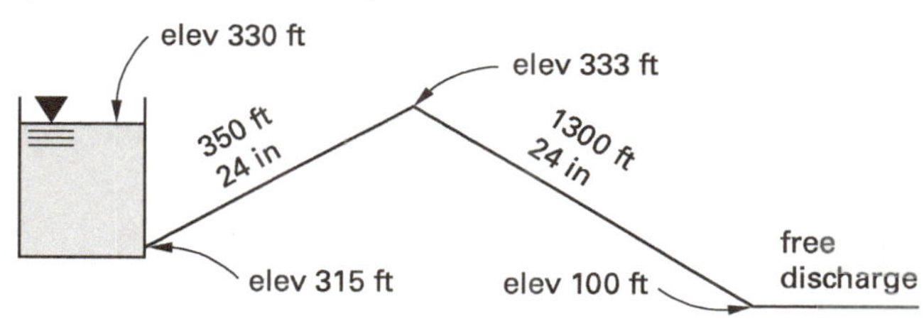

(not to scale)

(A) 6.5 psi, −28 psi

(B) 39 psi, −8.0 psi

(C) 93 psi, 7.8 psi

(D) 100 psi, 0.0 psi

31. An ideal compressible fluid is flowing from point A to point B. The properties of the fluid at points A and B are as follows.

	A	B
density	0.381 lbm/ft³	0.0508 lbm/ft³
internal energy	119.1 Btu/lbm	53.1 Btu/lbm
pressure	98.4 psia	5.85 psia

The change in enthalpy between points A and B is most nearly

(A) −93 Btu/lbm (decrease)

(B) −66 Btu/lbm (decrease)

(C) −27 Btu/lbm (decrease)

(D) 77 Btu/lbm (increase)

32. A 3 in diameter steel sphere achieves a terminal velocity of 122 in/sec when dropping through a tall column of liquid. The density of the liquid is 87 lbm/ft^3. The density of steel is 488 lbm/ft^3. The final drag coefficient of the sphere is most nearly

(A) 0.22

(B) 0.34

(C) 0.40

(D) 0.48

33. Water in a supply tank is used to form a small fountain of water as shown. Forty feet of 4 in schedule-40 steel pipe containing five long-radius 90° elbows and one gate valve run from the bottom of the tank to the discharge point. The opening in the tank has a sharp edge. If the minimum height that the 50°F water must achieve is 4 ft, the water depth in the tank should never be less than

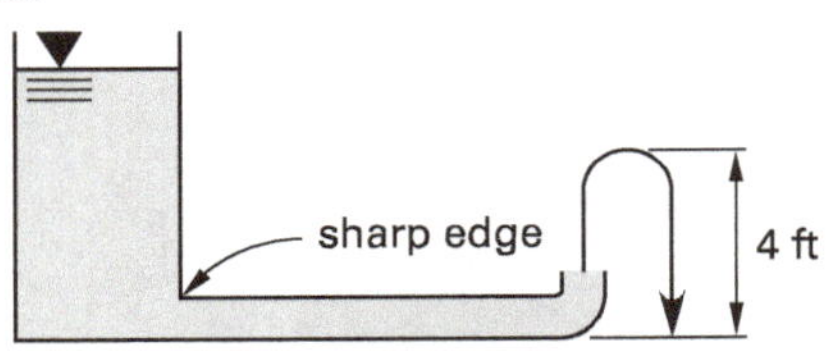

(not to scale)

(A) 15 ft

(B) 21 ft

(C) 33 ft

(D) 54 ft

34. A feedwater pump delivers 100 gpm water to a system. Because of concerns regarding the potential for erosion at the elbows, it has been specified that the water velocity in the pipe must not exceed 7.0 ft/sec. What minimum size schedule-40 pipe (nominal) will carry the flow without exceeding the maximum velocity criteria?

(A) 2 in

(B) 2½ in

(C) 4 in

(D) 20 in

35. A two-cylinder Otto cycle internal combustion engine produces 33 hp at 400 rpm. The bore is 8 in, and the stroke is 8 in. What is most nearly the brake mean effective pressure?

(A) 3.4 psig

(B) 21 psig

(C) 41 psig

(D) 81 psig

36. A hydraulic wood splitter with a double-acting cylinder operating at 2500 psi has a 24 in stroke in both directions. The hydraulic oil at normal operating temperature has a specific gravity of 0.85. The wood splitter has a 3 in bore and a 1 in diameter cylinder rod. The maximum retraction time for the cylinder rod is 5 sec. A four-way valve with a valve flow coefficient of 2.9 is used for cycling the wood splitter. The pressure drop across the four-way valve is most nearly

(A) 6.2 psi

(B) 6.7 psi

(C) 7.3 psi

(D) 7.9 psi

37. A heat pump supplies 900,000 Btu/day to a 3500 ft^2 home in order to maintain the temperature inside at 78°F when the temperature outside is 30°F. Electricity costs 3.7 cents/kW-hr. The minimum cost per 30 day month to operate the heat pump is most nearly

(A) $18

(B) $26

(C) $41

(D) $89

38. Enriched uranium (with a thermal conductivity of 20 Btu/hr-ft-°F) fuel plates 0.25 in thick are used in a nuclear reactor core. The coolant maintains one side of a plate at the edge of the core at 500°F and the other side at 750°F. If the internal heat generation of the plate is a uniform 9×10^7 Btu/hr-ft^3, the maximum temperature in the plate is most nearly

(A) 580°F

(B) 890°F

(C) 1700°F

(D) 1800°F

39. Steam at 900°F enters a turbine that operates in a Rankine cycle. If the cycle operates between pressures of 1000 psia and 20 psia, the efficiency of the cycle is most nearly

(A) 10%

(B) 20%

(C) 30%

(D) 50%

40. Water temperature is increased in a heat exchanger at the expense of a decrease in temperature of a non-foaming oil. The following information is known.

number of tube passes	2
tube (inside) area	40 ft^2
overall heat transfer coefficient, based on inside area	280 Btu/hr-ft^2-°F
tube fluid	water being heated
tube fluid temperature in	60°F
tube fluid flow rate	12,000 lbm/hr
number of shell passes	1
shell fluid	oil being cooled
shell fluid temperature in	280°F
shell fluid specific heat	0.60 Btu/lbm-°F
shell fluid flow rate	20,000 lbm/hr

The temperatures of the oil and the water, respectively, as they leave the heat exchanger are most nearly

(A) 120°F, 160°F

(B) 120°F, 230°F

(C) 180°F, 160°F

(D) 180°F, 230°F

41. 4 lbm/min of steam at 200 psia and 600°F are mixed with 11 lbm/min of steam at 200 psia and 70% quality in an adiabatic steady-flow device. The quality of the outlet mixture is most nearly

(A) 71%

(B) 82%

(C) 91%

(D) 97%

42. The rectangular wall of a furnace consists of 3 in fire-clay brick surrounded by 0.25 in of steel on the outside. There are six 0.25 in diameter mild steel bolts per square foot connecting the steel to the brick and extending completely through both the steel and the brick. The furnace is surrounded by 70°F air (with a film coefficient of 1.65 Btu/hr-ft^2-°F), while the inner surface of the brick is maintained at a constant 1000°F. The heat flux per square foot through this wall is most nearly

(A) 160 Btu/hr

(B) 300 Btu/hr

(C) 450 Btu/hr

(D) 930 Btu/hr

43. A 200 ton hydraulic press has a 90% efficient, 1200 rpm electric motor that is direct coupled to an 85% efficient hydraulic pump. The motor has a power factor of 0.85. The pump has a rated displacement of 1.93. The system's hydraulic pressure is 2500 psig. The press is powered by a 208 V three-phase electrical service. The current drawn by the motor is most nearly

(A) 30 A

(B) 50 A

(C) 60 A

(D) 80 A

44. A 40 W incandescent light bulb emitting 1500 lm has a surface temperature of 260°F. Still air at 80°F surrounds the bulb, and the bulb is modeled as a 2 in diameter sphere. The heat lost to free convection is most nearly

(A) 0.50 W

(B) 6.9 W

(C) 16 W

(D) 24 W

45. A 240 V electric jockey pump adds 300 ft of pressure head to a 60°F, 1.5 lbm/sec water flow. At operating speed, the motor is 95% efficient, and the pump is 65% efficient. What is most nearly the motor current draw?

(A) 2.5 A

(B) 4.1 A

(C) 6.2 A

(D) 11 A

46. The supply side of a closed-loop geothermal home heating system is plumbed with 400 ft of 1 in schedule-40 PVC pipe. The pipe is buried 3 ft underground and runs from the house to the bottom of a pond. The portion of the pipe that is underground is insulated with a 0.125 in thick layer of insulation ($k = 0.03$ Btu/hr-ft-°F). The glycol solution traveling through the 250 ft of pipe on the bottom of the pond is heated to 56°F by the liquid water at the bottom of the pond. The temperature of the soil surrounding the pipe is 20°F. The soil is saturated and frozen and has a thermal conductivity of 1.5 Btu/hr-ft-°F. The maximum heat loss per unit length of pipe experienced anywhere in the loop is most nearly

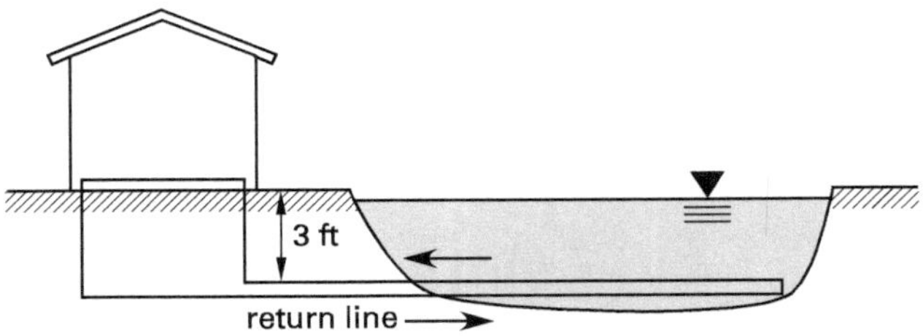

(not to scale)

(A) 0.50 Btu/hr-ft

(B) 1.5 Btu/hr-ft

(C) 13 Btu/hr-ft

(D) 26 Btu/hr-ft

47. A steel rod with a circular cross section has both ends built in between two walls. The rod is 11.5 ft long and has a diameter of 2.4 in. If the temperature is gradually increased, the rise in temperature at which the rod will first begin to buckle is most nearly

(A) 68°F

(B) 100°F

(C) 140°F

(D) 170°F

48. The thermocouple reading of a flame in a furnace is 550°F, and the temperature of the inside furnace walls is 400°F. The convection coefficient inside the furnace is 25 Btu/hr-ft^2-°F. The true temperature of the flame is most nearly

(A) 580°F

(B) 620°F

(C) 850°F

(D) 1000°F

49. For a liquid-to-liquid heat exchanger, what flow regime for the liquids makes heat transfer most efficient?

(A) supersonic

(B) turbulent

(C) transition

(D) laminar

50. A two-stage air compressor with an ideal intercooler takes air at 14.7 psia and 70°F and compresses it to 200 psia. The pressure staging is ideal. The mass flow rate of air is 10 lbm/min. The heat removed by the intercooler is most nearly

(A) 570 Btu/min

(B) 870 Btu/min

(C) 2400 Btu/min

(D) 4500 Btu/min

51. Water exits from an open holding tank to the atmosphere through a small opening 11 ft below the surface of the water. The volume of water in the tank remains constant. The ideal velocity of the water leaving the tank is most nearly

(A) 4.7 ft/sec

(B) 19 ft/sec

(C) 27 ft/sec

(D) 88 ft/sec

52. Atmospheric air flowing at 800 ft^3/min and at a wet-bulb temperature of 75°F and a dry-bulb temperature of 98°F is conditioned to a dry-bulb temperature of 67°F and 50% relative humidity in an evaporative cooler. The amount of heat removed by the cooling system and added by the heating system is most nearly

(A) heat removed = 1100 Btu/hr
heat added = 260 Btu/hr

(B) heat removed = 1600 Btu/hr
heat added = 3700 Btu/hr

(C) heat removed = 49,000 Btu/hr
heat added = 24,000 Btu/hr

(D) heat removed = 65,000 Btu/hr
heat added = 15,000 Btu/hr

53. Steam at 250 psia enters an isentropic turbine operating in a Rankine cycle. The steam exits the turbine at 100°F. The moisture content of the steam in the turbine is not to exceed 8%. The minimum temperature to which the high-pressure steam must be heated is most nearly

(A) 170°F

(B) 530°F

(C) 1000°F

(D) 1300°F

54. At the beginning of the compression stroke of an air-standard Otto cycle, the pressure is 14.7 psia and the temperature is 68°F. The maximum pressure in the cycle is 510 psia, and the compression ratio is 8.5:1. The difference in efficiency between this cycle and a Carnot cycle operating between the same temperatures is most nearly

(A) 10%

(B) 20%

(C) 30%

(D) 40%

55. The net output power from a steam-based Rankine cycle is 250 MW. The cycle has steam entering an isentropic turbine at 1100°F and 1250 psia, and cooling water entering the condenser at 56°F and leaving as saturated liquid at 93°F. The condenser pressure is 5 psia. The mass flow rate of steam is most nearly

(A) 2.9×10^4 lbm/hr

(B) 3.3×10^5 lbm/hr

(C) 4.7×10^5 lbm/hr

(D) 1.6×10^6 lbm/hr

56. A pressure control valve for limiting hydraulic system pressure is redesigned to use a helical spring with squared and ground ends. The spring will be made from 0.10 in diameter wire having a shear modulus of 11.5×10^6 psi. The maximum system operating pressure is to be 250 psi. The effective control area where the spring ultimately acts to limit system pressure is 0.123 in^2. The maximum spring deflection is to be $^7/_{16}$ in. The spring index is 8. The spring must have

(A) 2 coils

(B) 4 coils

(C) 6 coils

(D) 8 coils

57. A regenerative vapor power cycle with one open feedwater heater operates between pressures of 600 psia and 10 psia with a turbine inlet steam temperature of 860°F. Steam at 130 psia is extracted and sent to the open feedwater heater. The percentage of the total flow passing through the second-stage isentropic turbine is most nearly

(A) 14%

(B) 51%

(C) 75%

(D) 86%

58. Two metal blocks are brought into contact with each other and allowed to reach thermal equilibrium. One has a mass of 60 lbm, is made of copper, and is at 50°F. The other has a mass of 30 lbm, is made of stainless steel, and is at 420°F. If the blocks are considered as a system and losses to the surroundings are neglected, the final temperature is most nearly

(A) 170°F

(B) 190°F

(C) 230°F

(D) 260°F

59. A natural gas has the following molar analysis.

CH_4	80.32%
C_2H_6	5.75%
C_3H_8	1.79%
C_4H_{10}	1.69%
N_2	10.45%

When the gas is burned in air, the following dry analysis describes the products.

CO_2	7.9%
CO	0.1%
O_2	7%
N_2	85%

The air/fuel ratio is most nearly

(A) 4.8 kmol air/kmol fuel

(B) 7.2 kmol air/kmol fuel

(C) 11 kmol air/kmol fuel

(D) 14 kmol air/kmol fuel

60. Water flows through a cast-iron pipe at 1800 gpm. The inside diameter of the pipe is 10 in. The Darcy friction factor is 0.0195. The pressure drop over an 185 ft length of pipe is most nearly

(A) 1.6 psi

(B) 3.6 psi

(C) 16 psi

(D) 19 psi

61. Steam at 300°F flows in an insulated pipe with an inside diameter of 4 in and an outside diameter of 4.5 in. The pipe is mild steel with a thermal conductivity of 31 Btu-ft/hr-ft^2-°F. The pipe is covered with a 1 in thickness of 85% magnesia with a thermal conductivity of 0.03 Btu-ft/hr-ft^2-°F. The film coefficient for the contact between the pipe and the insulation is 15 Btu/hr-ft^2-°F, and the outside film coefficient is 2 Btu/hr-ft^2-°F. The ambient temperature is 65°F. Most nearly, the radial conduction through the pipe wall per foot of pipe is

(A) 60 Btu/hr-ft

(B) 87 Btu/hr-ft

(C) 100 Btu/hr-ft

(D) 190 Btu/hr-ft

62. The pumping curve shown is for a variable speed 3450 rpm water pump. The system (or operating) curve for an application that will use this pump is superimposed on the pumping curve. At maximum load, 500 gpm at 85 ft of head is needed.

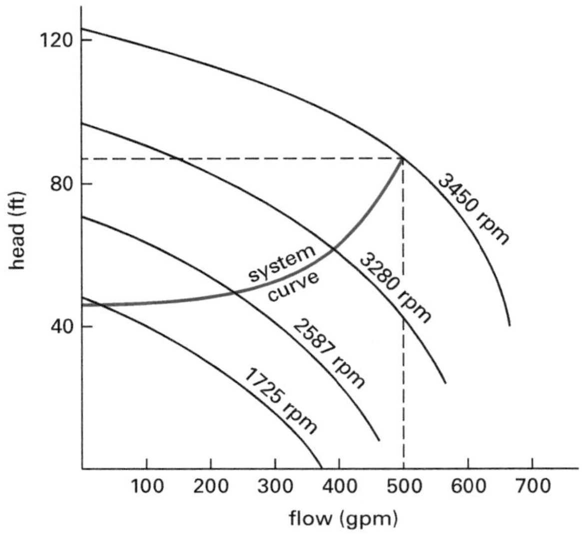

During an off-peak condition, only one-half the design flow is needed. Most nearly, the pump speed should then be reduced to

(A) 1400 rpm

(B) 1700 rpm

(C) 2600 rpm

(D) 3300 rpm

63. A gas pipeline is being added near a small residential neighborhood just outside of a town. The proposed pipeline is 1 mi long and will carry gas at 30 psig and 300°F. There are 25 homes within $\frac{1}{8}$ mi of the proposed pipeline, all homes are one or two stories, and there are no plans for significant further development. What ASME location class is most appropriate for this area?

(A) class 1

(B) class 2

(C) class 3

(D) class 4

64. A six-cylinder, four-stroke engine has a 100.4 mm bore and an 84.4 mm stroke. The brake mean effective pressure while running at 4200 rpm is 120 psig. The brake horsepower is most nearly

(A) 100 hp

(B) 160 hp

(C) 180 hp

(D) 210 hp

65. A round duct industrial ventilation system is shown. The static pressure at the branch is 0.10 in wg, and the exit pressure at each diffuser is 0.03 in wg. The straight branch is 20 ft long, and the take-off branch is 30 ft long. The airflow rate needed in each branch is 1000 ft^3/min. Dynamic losses are negligible.

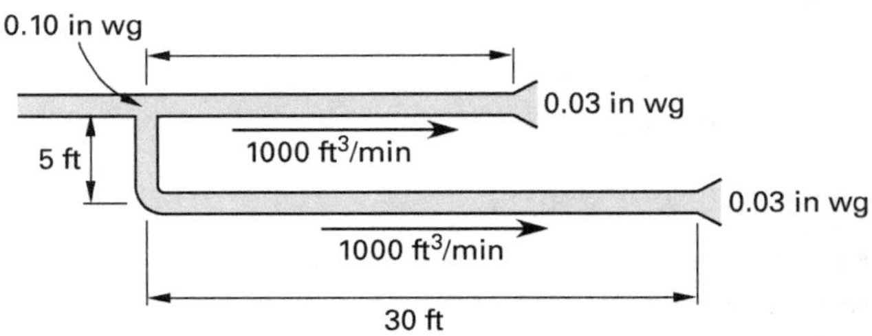

(not to scale)

Most nearly, the minimum diameters needed for the straight branch and the take-off branch, respectively, are

(A) 9 in and 12 in

(B) 9 in and 14 in

(C) 11 in and 12 in

(D) 11 in and 14 in

66. What is the purpose of the three-way valve shown?

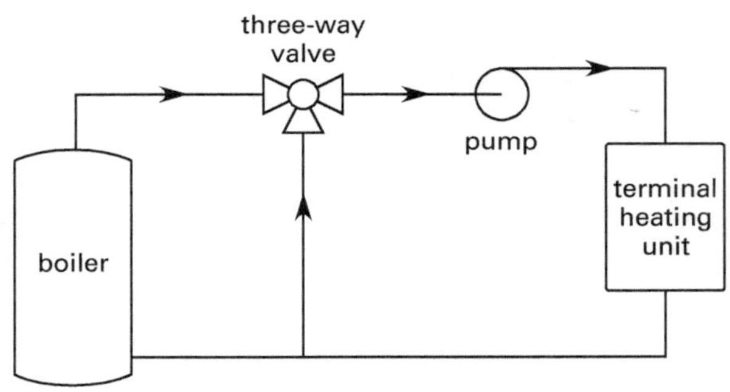

(A) modulate flow volume through the boiler

(B) modulate flow volume through the pump

(C) mix the heating water to the desired temperature

(D) maintain the constant pressure drop needed by the pump

67. A 110 hp internal combustion engine burns propane at a rate of 0.01 lbm/sec when mixed with theoretical air. The engine takes in air and fuel at 77°F and 1 atm. The air-fuel mixture burns completely and leaves the engine at 1240°F. The rate of heat transfer from the engine and exhaust is most nearly

(A) −140 Btu/sec

(B) −93 Btu/sec

(C) −68 Btu/sec

(D) 33 Btu/sec

68. Octane is burned completely with theoretical air at atmospheric pressure. The dew point of the products is most nearly

(A) 130°F

(B) 150°F

(C) 200°F

(D) 210°F

69. A schematic diagram of a crossflow, air-to-air, flat plate heat exchanger is shown. This type of heat exchanger will transfer heat between a cold airstream and a warm airstream. Quadrants of the heat exchanger plate are indicated.

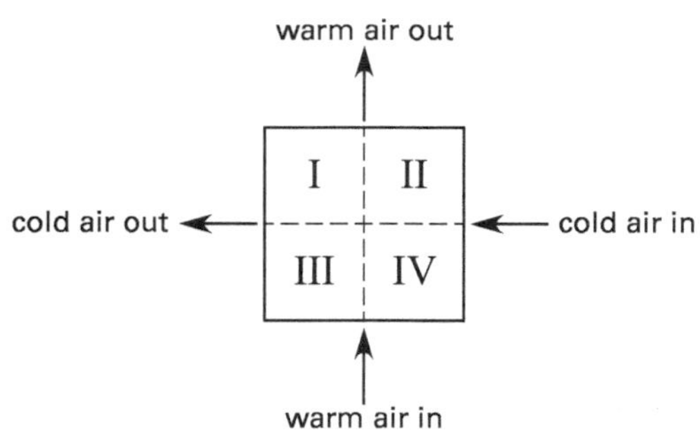

Which quadrant has the LOWEST rate of heat transfer?

(A) I

(B) II

(C) III

(D) IV

70. An engineering consulting company is considering a $100,000 investment in new computers and software. The software and equipment have only a five-year lifetime. The expected annual operating costs will be $20,000/yr for the first two years and $25,000/yr for the remaining period. The company would like to get a 15% return on its investment. Most nearly, the additional annual profit needed to realize this goal is

(A) $37,000

(B) $42,000

(C) $47,000

(D) $52,000

71. A fan blows 30,000 cfm of air into a 24 in × 36 in duct. Most nearly, the 100% effective duct length for this duct is

(A) 5.4 ft

(B) 8.6 ft

(C) 11 ft

(D) 14 ft

72. An air compressor takes in atmospheric air at 0 psig and 60°F. The compressor compresses the air into a tank that is at 150 psig. The compression temperature is 300°F. Most nearly, the polytropic index for this process is

(A) 0.83

(B) 0.99

(C) 1.2

(D) 1.4

73. Steam at 400°F and 2000 psia enters a steam turbine and is expanded to 200 psia. The turbine has an isentropic efficiency of 75%. Most nearly, the enthalpy of the steam as it exits the turbine is

(A) 980 Btu/lbm

(B) 1100 Btu/lbm

(C) 1300 Btu/lbm

(D) 1500 Btu/lbm

74. A simple gas turbine cycle on the front half of a combined cycle intakes atmospheric air at 60°F and 14.7 psia. The pressure ratio is 5:1 and the air-fuel ratio is 150. The heating value of the fuel is 20,000 Btu/lbm, and the combustion efficiency is 80%. Most nearly, the enthalpy of the air as it enters the turbine is

(A) 330 Btu/lbm

(B) 360 Btu/lbm

(C) 390 Btu/lbm

(D) 430 Btu/lbm

75. In a cogeneration plant at a sugar factory, 100 kW of electricity is generated by a steam cycle. The heat rejected in the condenser of the steam cycle is used to power a cane juice boiler. The heating value of the fuel is 4000 Btu/lbm, and 250 lbm/hr of fuel is fed to the steam boiler. Most nearly, the rate of heat available to the cane juice boiler is

(A) 660,000 Btu/hr

(B) 720,000 Btu/hr

(C) 800,000 Btu/hr

(D) 890,000 Btu/hr

76. In an ideal combined cycle power plant (gas turbine cycle plus steam Rankine cycle), the ideal gas turbine cycle is 45% efficient and the ideal steam cycle is 30% efficient. Most nearly, what is the overall efficiency of the combined cycle?

(A) 14%

(B) 29%

(C) 62%

(D) 75%

77. In a food processing drying oven, 2000 cfm of atmospheric air at 65°F and 40% relative humidity is filtered and heated to 250°F. The air then passes over the food product, which goes through the dryer at 12,000 lbm/hr. The air exits the dryer at 140°F and 40% relative humidity. The food product has a specific heat of 0.75 Btu/lbm-°F. The latent heat of water is 940 Btu/lbm. Most nearly, the temperature rise of the food product as it goes through the dryer is

(A) 18°F

(B) 35°F

(C) 73°F

(D) 120°F

78. Which air waste heat recovery system has the potential to recover both latent and sensible heat?

(A) run-around

(B) heat pump

(C) heat recovery wheel

(D) heat pipe

79. A plant manager is considering replacing an air compressor system with a new 110 hp belt drive model. The compressor will operate at full load for 1800 hr/yr. The new model has an initial cost of $80,000, an annual maintenance cost of $2000, and a salvage value of $10,000 after an expected useful life of 10 yr. The expected interest rate is 10%, and the cost of electricity is $0.12/kW-hr. Most nearly, the equivalent uniform annual cost for this compressor system is

(A) $23,000

(B) $26,000

(C) $29,000

(D) $32,000

80. A heat recovery wheel is placed in a system between exhaust air and makeup air. The wheel is 40% efficient and can recover only sensible heat. Air is exhausted at 75°F dry-bulb and 50% relative humidity, and makeup air enters at 35°F dry-bulb and 20% relative humidity. The rates of the exhaust and make-up airstreams are equal. Most nearly, the amount of heat that can be recovered is

(A) 3 Btu/lbm

(B) 4 Btu/lbm

(C) 5 Btu/lbm

(D) 6 Btu/lbm

Answer Key

No.	Answer	No.	Answer	No.	Answer
1.	C	28.	B	55.	D
2.	C	29.	B	56.	C
3.	B	30.	A	57.	D
4.	B	31.	A	58.	B
5.	D	32.	D	59.	D
6.	C	33.	B	60.	A
7.	C	34.	B	61.	A
8.	B	35.	D	62.	C
9.	D	36.	A	63.	B
10.	C	37.	B	64.	B
11.	B	38.	B	65.	C
12.	B	39.	C	66.	C
13.	D	40.	C	67.	C
14.	C	41.	B	68.	A
15.	D	42.	D	69.	A
16.	B	43.	B	70.	D
17.	C	44.	B	71.	D
18.	B	45.	B	72.	C
19.	A	46.	D	73.	C
20.	C	47.	A	74.	A
21.	B	48.	A	75.	A
22.	C	49.	B	76.	C
23.	C	50.	A	77.	C
24.	A	51.	C	78.	C
25.	D	52.	D	79.	D
26.	B	53.	C	80.	B
27.	B	54.	B		

Solutions
Practice Exam

1. PVC, polystyrene, and ABS are examples of amorphous polymers, while polypropylene is a crystalline polymer. In general, crystalline polymers have better chemical resistance than amorphous polymers.

The answer is (C).

2. The total dynamic head provided by the pump, h_{pump}, is that required to overcome the pressure of the pressurized tank, the friction and fitting losses, h_{losses}, and the static discharge head to the pressurized tank waterline, less the positive static head of the water column to the storage tank above.

Bernoulli's equation describes the energy (as head) relationships as

$$h_{\text{pump}} + \frac{p_1}{\gamma_1} + \frac{v_1^2}{2g} + z_1 = \frac{p_2}{\gamma_2} + \frac{v_2^2}{2g} + z_2 + h_{\text{losses}}$$

The equation can be rearranged to solve for the required pump head, choosing the water surface in the open tank as point 1 and the water surface of the pressurized tank as point 2. The velocities v_1 and v_2 are approximately zero.

$$\begin{aligned} h_{\text{pump}} &= \frac{p_2 - p_1}{\gamma} + \frac{v_2^2 - v_1^2}{2g} + z_2 - z_1 + h_{\text{losses}} \\ &= \frac{\left(15\ \frac{\text{lbf}}{\text{in}^2} - 0\ \frac{\text{lbf}}{\text{in}^2}\right)\left(12\ \frac{\text{in}}{\text{ft}}\right)^2}{62.4\ \frac{\text{lbf}}{\text{ft}^3}} \\ &\quad +7\ \text{ft water} - 15\ \text{ft water} + 12\ \text{ft water} \\ &= 38.62\ \text{ft water} \quad (39\ \text{ft water}) \end{aligned}$$

The answer is (C).

3. The heat added by the compressor comes from the work done by the compressor. The COP of the process is given as 4.4. The COP is the refrigeration effect divided by the power.

$$\text{COP} = \frac{\dot{q}_{\text{entering}}}{P_{\text{in}}}$$

$$P_{\text{in}} = \frac{\dot{q}_{\text{entering}}}{\text{COP}}$$

The capacity required by the evaporator to cool the water is

$$\begin{aligned} \dot{q}_{\text{evaporator}} &= \dot{m}c_p\Delta T \\ &= \dot{V}\rho c_p(T_{\text{entering}} - T_{\text{leaving}}) \\ &= \frac{\left(\frac{640\ \frac{\text{gal}}{\text{min}}}{7.48\ \frac{\text{gal}}{\text{ft}^3}}\right)\left(60\ \frac{\text{min}}{\text{hr}}\right) \times \left(62.4\ \frac{\text{lbm}}{\text{ft}^3}\right)\left(1.0\ \frac{\text{Btu}}{\text{lbm-°F}}\right) \times (55°\text{F} - 43°\text{F})}{12{,}000\ \frac{\text{Btu}}{\text{ton-hr}}} \\ &= 320\ \text{tons} \end{aligned}$$

The compressor power is

$$\begin{aligned} P_{\text{in}} &= \frac{\dot{q}_{\text{evaporator}}}{\text{COP}} \\ &= \frac{320\ \text{tons}}{4.4} \\ &= 72.7\ \text{tons} \quad (73\ \text{tons}) \end{aligned}$$

The answer is (B).

4. In a principal, cavalier, or sectional view, at least one projection (also called a projection plane) is parallel with the plane of the screen or paper (or in other words, perpendicular to the direction from which the object is being viewed). The view shown is isometric.

The answer is (B).

5. The tangential velocity at a radius of 1.5 ft is

$$v_t = \omega r = \left(188.5\ \frac{\text{rad}}{\text{sec}}\right)(1.5\ \text{ft}) = 282.75\ \text{ft/sec}$$

The air is introduced at three times the tangential velocity, so the velocity of the introduced air is

$$v_{\text{air}} = 3v_t = (3)\left(282.75\ \frac{\text{ft}}{\text{sec}}\right) = 848.25\ \text{ft/sec}$$

The velocity of sound is

$$a = \sqrt{kg_cRT}$$

Air temperature is not given, however, so an approximate average value must be assumed. At normal ambient conditions, the speed of sound is about 1125 ft/sec.

The Mach number is

$$\mathrm{M} = \frac{\mathrm{v}}{a} = \frac{848.25\ \frac{\mathrm{ft}}{\mathrm{sec}}}{1125\ \frac{\mathrm{ft}}{\mathrm{sec}}} = 0.754 \quad (0.75)$$

The answer is (D).

6. Because the relief valve discharges into atmospheric pressure and the hydrostatic head is present in both the pump and relief lines, the total pressure drop across the valve is the oil pressure.

Valve flow rate is related to the pressure drop across the valve through the following formula (which is not dimensionally consistent). C_v is the valve flow coefficient.

$$\begin{aligned} Q_{\mathrm{gpm}} &= C_v\sqrt{\frac{\Delta p_{\mathrm{psi}}}{\mathrm{SG}}} \\ &= 0.53\sqrt{\frac{5.775\ \frac{\mathrm{lbf}}{\mathrm{in}^2}}{0.90}} \\ &= 1.34\ \mathrm{gpm} \quad (1.3\ \mathrm{gpm}) \end{aligned}$$

The answer is (C).

7. A liquid's kinematic viscosity in Saybolt universal seconds is the time it takes for 60 cm^3 of the liquid to flow through a calibrated tube at a controlled temperature. Intuitively, then, neither 0.90 SUS nor 32,000 SUS is a plausible answer choice for SAE 10W-30 oil.

For viscosities greater than about 240 SUS, one centistoke is equal to about 4.65 SUS. The kinematic viscosity in Saybolt universal seconds is

$$\begin{aligned} \nu &= (110\ \mathrm{cSt})\left(4.65\ \frac{\mathrm{SUS}}{\mathrm{cSt}}\right) \\ &= 512\ \mathrm{SUS} \quad (510\ \mathrm{SUS}) \end{aligned}$$

The answer is (C).

8. At $T = 40°\mathrm{C}$, the saturation pressure of water vapor is 7.38 kPa.

Determine the specific humidity of the air at the inlet and outlet. The specific humidity, or mass ratio of water vapor to dry air, is given by

$$\omega = \frac{m_w}{m_a}$$

The specific humidity is expressed in units of kg/kg or as a dimensionless quantity. Under the ideal gas law, the specific humidity can be written as follows.

$$\begin{aligned} \omega &= \frac{m_w}{m_a} = \frac{\frac{p_wV}{R_wT}}{\frac{p_aV}{R_aT}} = \frac{0.622p_w}{p_a} \\ &= \frac{0.622p_w}{p - p_w} \end{aligned}$$

Initially the air is dry, so $\omega_1 = 0$. The air exiting the channel at point 3 is saturated at 40°C. Therefore, the specific humidity at point 3 is

$$\omega_3 = \frac{0.622p_w}{p - p_w} = \frac{(0.622)(7.38\ \mathrm{kPa})}{100\ \mathrm{kPa} - 7.38\ \mathrm{kPa}} = 0.049$$

Since the channel is insulated, it can be treated as a two-inlet, one-outlet adiabatic, steady-flow system. Apply conservation of mass to both the dry air flow and water vapor flow through the channel. The mass of air entering equals the mass leaving.

$$\dot{m}_{a,1} = \dot{m}_{a,3}$$

The increase in water vapor is due to evaporation from the pool.

$$\dot{m}_{w,1} + \dot{m}_{\mathrm{evap}} = \dot{m}_{w,3}$$

Combining these relations yields

$$\begin{aligned} \dot{m}_{\mathrm{evap}} &= \dot{m}_{w,3} - \dot{m}_{w1} = \dot{m}_{a,3}\omega_3 - \dot{m}_{a,1}\omega_1 = \dot{m}_a(\omega_3 - \omega_1) \\ &= \dot{m}_a\omega_3 \quad [\text{since } \omega_1 = 0] \\ &= \left(12\ \frac{\mathrm{kg}}{\mathrm{s}}\right)(0.049) \\ &= 0.588\ \mathrm{kg/s} \quad (0.6\ \mathrm{kg/s}) \end{aligned}$$

The answer is (B).

9. The dimensions of 2½ in schedule-40 steel pipe, and the densities of steel and water are

$$D = 2.875 \text{ in}$$
$$d = 2.469 \text{ in}$$
$$A = 4.788 \text{ in}^2$$
$$\gamma_{\text{steel}} = \frac{490 \ \frac{\text{lbf}}{\text{ft}^3}}{\left(12 \ \frac{\text{in}}{\text{ft}}\right)^3} = 0.28 \text{ lbf/in}^3$$
$$\gamma_{\text{water}} = \frac{62.4 \ \frac{\text{lbf}}{\text{ft}^3}}{\left(12 \ \frac{\text{in}}{\text{ft}}\right)^3} = 0.0361 \text{ lbf/in}^3$$

Calculate the weights of pipe and water and the weight, w, per unit length.

$$\begin{aligned} w_{\text{pipe}} &= \frac{W_{\text{pipe}}}{L} = \gamma_{\text{steel}}\left(\frac{\pi}{4}(D^2 - d^2)\right) \\ &= \left(0.28 \ \frac{\text{lbf}}{\text{in}^3}\right)\left(\frac{\pi}{4}\left(\begin{array}{l}(2.875 \text{ in})^2 \\ \quad -(2.469 \text{ in})^2\end{array}\right)\right) \\ &= 0.477 \text{ lbf/in} \end{aligned}$$

$$\begin{aligned} w_{\text{water}} &= \frac{W_{\text{water}}}{L} = \gamma_{\text{water}} A = \left(0.0361 \ \frac{\text{lbf}}{\text{in}^3}\right)(4.788 \text{ in}^2) \\ &= 0.173 \text{ lbf/in} \end{aligned}$$

$$\begin{aligned} w &= w_{\text{pipe}} + w_{\text{water}} = 0.477 \ \frac{\text{lbf}}{\text{in}} + 0.173 \ \frac{\text{lbf}}{\text{in}} \\ &= 0.65 \text{ lbf/in} \end{aligned}$$

Calculate the moment of inertia for bending.

$$\begin{aligned} I &= \frac{\pi}{64}(D^4 - d^4) \\ &= \left(\frac{\pi}{64}\right)((2.875 \text{ in})^4 - (2.469 \text{ in})^4) \\ &= 1.53 \text{ in}^4 \end{aligned}$$

Calculate the deflection for a simply supported beam carrying a uniformly distributed load. The modulus of elasticity for steel is $E = 30 \times 10^6 \text{ lbf/in}^2$.

$$\begin{aligned} y_{\max} &= -\frac{5wL^4}{384EI} \\ &= -\frac{(5)\left(0.65 \ \frac{\text{lbf}}{\text{in}}\right)\left((20 \text{ ft})\left(12 \ \frac{\text{in}}{\text{ft}}\right)\right)^4}{(384)\left(30 \times 10^6 \ \frac{\text{lbf}}{\text{in}^2}\right)(1.53 \text{ in}^4)} \\ &= 0.611 \text{ in} \quad (0.61 \text{ in}) \end{aligned}$$

The answer is (D).

10. For a steady, laminar, incompressible flow, the pressure drop is

$$\begin{aligned} \Delta p &= \frac{128\mu \dot{V} L}{\pi d^4} \\ &= \frac{(128)\left(26.37 \times 10^{-6} \ \frac{\text{lbf-sec}}{\text{ft}^2}\right) \times \left(3 \times 10^{-5} \ \frac{\text{ft}^3}{\text{sec}}\right)(3.4 \text{ ft})\left(12 \ \frac{\text{in}}{\text{ft}}\right)^4}{\pi(0.018 \text{ in})^4} \\ &= 21{,}647 \text{ lbf/ft}^2 \quad (22{,}000 \text{ lbf/ft}^2) \end{aligned}$$

The answer is (C).

11. The hoop stress, or circumferential stress, is

$$\sigma_h = \frac{pr}{t}$$

The longitudinal stress is half this value.

$$\sigma_l = \frac{pr}{2t}$$

The radius of the tank is

$$\begin{aligned} r &= \frac{d}{2} \\ &= \frac{300 \text{ mm}}{2} \\ &= 150 \text{ mm} \end{aligned}$$

The principal stresses are the hoop and longitudinal stresses, and there is no torsion, τ, on the tank, so the stress at an orientation of 55° is

$$\begin{aligned}\sigma_\theta &= \frac{\sigma_h+\sigma_l}{2}+\left(\frac{\sigma_h-\sigma_l}{2}\right)\cos 2\theta+\tau\sin 2\theta\\ &= \frac{\frac{pr}{t}+\frac{pr}{2t}}{2}+\left(\frac{\frac{pr}{t}-\frac{pr}{2t}}{2}\right)\cos 2\theta+\tau\sin 2\theta\\ &= \frac{3pr}{4t}+\left(\frac{pr}{4t}\right)\cos 2\theta+\tau\sin 2\theta\\ \sigma_{55^\circ} &= \frac{3pr}{4t}+\left(\frac{pr}{4t}\right)\cos(2)(55^\circ)+(0\text{ Pa})\sin(2)(55^\circ)\\ &= 0.665pr/t\end{aligned}$$

The alternating layers are oriented 90° from each other, since the 55° orientation is referenced to both longitudinal and circumferential directions. Check the stress at $90^\circ - 55^\circ = 35^\circ$.

$$\sigma_{35^\circ} = \frac{3pr}{4t}+\left(\frac{pr}{4t}\right)\cos(2)(35^\circ) = 0.836pr/t$$

This is a higher stress than in the other direction.

The thickness, t, that is needed to limit the stress, σ_{55°, to 115 MPa is

$$\begin{aligned}t &= \frac{0.836pr}{\sigma_{55^\circ}}\\ &= \frac{(0.836)(10\text{ MPa})(150\text{ mm})}{115\text{ MPa}}\\ &= 10.9\text{ mm}\quad(11\text{ mm})\end{aligned}$$

The answer is (B).

12. The interest rate is unknown.

$$\begin{aligned}D_j &= \frac{C-S_n}{n} = \frac{\$525{,}000-\$0}{15} = \$35{,}000\\ \text{BV}_7 &= C-jD_j = \$525{,}000-(7)(\$35{,}000) = \$280{,}000\end{aligned}$$

Gain on the sale of a depreciated asset is

$$\text{gain} = S-\text{BV}_7 = \$700{,}000-\$280{,}000 = \$420{,}000$$

$$\begin{aligned}P = &-\$525{,}000+(\$700{,}000)(P/F,i\%,7)\\ &+(\$35{,}000)(P/A,i\%,7)(0.40)\\ &-(\$420{,}000)(P/F,i\%,7)(0.40)\\ &-(\$25{,}000)(P/A,i\%,7)(1-0.40)\\ &+(\$45{,}000)(P/A,i\%,3)(1-0.40)\\ &+(\$75{,}000)(P/A,i\%,7)(1-0.40)\\ &-(\$75{,}000)(P/A,i\%,3)(1-0.40)\end{aligned}$$

Neglect the benefit of a favorable capital gains tax rate on the gain above \$525,000.

Simplify and combine terms.

$$\begin{aligned}P &= -\$525{,}000\\ &\quad+\left(\begin{array}{l}\$700{,}000\\ \quad-(\$420{,}000)(0.40)\end{array}\right)(P/F,i\%,7)\\ &\quad+\left(\begin{array}{l}(\$35{,}000)(0.40)\\ \quad-(\$25{,}000)(0.60)\\ \quad+(\$75{,}000)(0.60)\end{array}\right)(P/A,i\%,7)\\ &\quad+\left(\begin{array}{l}(\$45{,}000)(0.60)\\ \quad-(\$75{,}000)(0.60)\end{array}\right)(P/A,i\%,3)\\ &= -\$525{,}000+(\$532{,}000)(P/F,i\%,7)\\ &\quad+(\$44{,}000)(P/A,i\%,7)\\ &\quad+(-\$18{,}000)(P/A,i\%,3)\end{aligned}$$

Try $i = 10\%$.

$$\begin{aligned}P &= -\$525{,}000+(\$532{,}000)(0.5132)\\ &\quad+(\$44{,}000)(4.8684)\\ &\quad+(-\$18{,}000)(2.4869)\\ &= -\$82{,}532\end{aligned}$$

Try $i = 5\%$.

$$\begin{aligned}P &= -\$525{,}000+(\$532{,}000)(0.7107)\\ &\quad+(\$44{,}000)(5.7864)\\ &\quad+(-\$18{,}000)(2.7232)\\ &= \$58{,}676\end{aligned}$$

Use linear interpolation.

$$\begin{aligned}\text{ROR} &\approx 5\%+(10\%-5\%)\left(\frac{\$58{,}676}{\$58{,}676-(-\$82{,}532)}\right)\\ &= 7.08\%\quad(7.1\%)\end{aligned}$$

The answer is (B).

13. The Bernoulli equation can be used to solve this problem.

$$\frac{p_1}{\gamma} + \frac{v_1^2}{2g} + z_1 + h_p = \frac{p_2}{\gamma} + \frac{v_2^2}{2g} + z_2 + h_L$$

$$\frac{\left(100\ \frac{\text{lbf}}{\text{in}^2}\right)\left(12\ \frac{\text{in}}{\text{ft}}\right)^2}{62.4\ \frac{\text{lbf}}{\text{ft}^3}} + \frac{v_1^2}{(2)\left(32.2\ \frac{\text{ft}}{\text{sec}^2}\right)} + 0\ \text{ft} + h_p$$
$$= \frac{\left(500\ \frac{\text{lbf}}{\text{in}^2}\right)\left(12\ \frac{\text{in}}{\text{ft}}\right)^2}{62.4\ \frac{\text{lbf}}{\text{ft}^3}} + \frac{v_2^2}{(2)\left(32.2\ \frac{\text{ft}}{\text{sec}^2}\right)} + 30\ \text{ft} + 10\ \text{ft}$$

The velocities in both the suction and discharge lines can be found using

$$\dot{V} = vA$$

The flow rate in each section of the pipe is 7000 gpm. The velocity in the suction line is

$$\frac{7000\ \frac{\text{gal}}{\text{min}}}{\left(7.48\ \frac{\text{gal}}{\text{ft}^3}\right)\left(60\ \frac{\text{sec}}{\text{min}}\right)} = v_1\left(\frac{\pi}{4}\right)\left(\frac{20\ \text{in}}{12\ \frac{\text{in}}{\text{ft}}}\right)^2$$
$$v_1 = 7.15\ \text{ft/sec}$$

The velocity in the discharge line is

$$\frac{7000\ \frac{\text{gal}}{\text{min}}}{\left(7.48\ \frac{\text{gal}}{\text{ft}^3}\right)\left(60\ \frac{\text{sec}}{\text{min}}\right)} = v_2\left(\frac{\pi}{4}\right)\left(\frac{12\ \text{in}}{12\ \frac{\text{in}}{\text{ft}}}\right)^2$$
$$v_2 = 19.9\ \text{ft/sec}$$

Substitute into the Bernoulli equation and solve for the required pump head.

$$\frac{\left(100\ \frac{\text{lbf}}{\text{in}^2}\right)\left(12\ \frac{\text{in}}{\text{ft}}\right)^2}{62.4\ \frac{\text{lbf}}{\text{ft}^3}} + \frac{\left(7.15\ \frac{\text{ft}}{\text{sec}}\right)^2}{(2)\left(32.2\ \frac{\text{ft}}{\text{sec}^2}\right)} + 0\ \text{ft} + h_p$$
$$= \frac{\left(500\ \frac{\text{lbf}}{\text{in}^2}\right)\left(12\ \frac{\text{in}}{\text{ft}}\right)^2}{62.4\ \frac{\text{lbf}}{\text{ft}^3}} + \frac{\left(19.9\ \frac{\text{ft}}{\text{sec}}\right)^2}{(2)\left(32.2\ \frac{\text{ft}}{\text{sec}^2}\right)} + 30\ \text{ft} + 10\ \text{ft}$$
$$h_p = 968\ \text{ft}$$

The required power of the pump is

$$P = \dot{V}\gamma h_p$$
$$= \frac{\left(7000\ \frac{\text{gal}}{\text{min}}\right)\left(62.4\ \frac{\text{lbf}}{\text{ft}^3}\right)(968\ \text{ft})}{\left(7.48\ \frac{\text{gal}}{\text{ft}^3}\right)\left(60\ \frac{\text{sec}}{\text{min}}\right)\left(550\ \frac{\text{ft-lbf}}{\text{hp-sec}}\right)}$$
$$= 1713\ \text{hp}\quad (1700\ \text{hp})$$

The answer is (D).

14. The additional power required can be determined using

$$P = F_D v$$

F_D is the drag force caused by the carrier, and v is the velocity of air relative to the car.

The drag force on the carrier is

$$F_D = \frac{\rho C_D A v^2}{2g_c}$$

The air density is approximately

$$\rho = 0.075\ \text{lbm/ft}^3$$

The carrier will be modeled as a flat plate. The projected area is

$$A = \frac{(35\ \text{in})(18\ \text{in})}{\left(12\ \frac{\text{in}}{\text{ft}}\right)^2} = 4.375\ \text{ft}^2$$

The relative wind speed is

$$\mathrm{v} = \frac{\left(65\ \frac{\text{mi}}{\text{hr}} + 10\ \frac{\text{mi}}{\text{hr}}\right)\left(5280\ \frac{\text{ft}}{\text{mi}}\right)}{3600\ \frac{\text{sec}}{\text{hr}}} = 110\ \text{ft/sec}$$

Determine the drag coefficient, C_D. Compute the Reynolds number.

$$\begin{aligned}\text{Re}_h &= \frac{\mathrm{v}h}{\nu} = \frac{\left(110\ \frac{\text{ft}}{\text{sec}}\right)(18\ \text{in})}{\left(1.58\times 10^{-4}\ \frac{\text{ft}^2}{\text{sec}}\right)\left(12\ \frac{\text{in}}{\text{ft}}\right)}\\ &= 1.04\times 10^6\end{aligned}$$

To use a drag coefficient chart, the width to height ratio is needed.

$$\frac{w}{h} = \frac{35\ \text{in}}{18\ \text{in}} = 1.94 \quad (2)$$

Using the Reynolds number and w/h ratio, the drag coefficient can be found graphically.

$$C_D \approx 1.18$$

The drag force is

$$\begin{aligned}F_D &= \frac{\rho C_D A \mathrm{v}^2}{2g_c}\\ &= \frac{\left(0.075\ \frac{\text{lbm}}{\text{ft}^3}\right)(1.18)(4.375\ \text{ft}^2)\left(110\ \frac{\text{ft}}{\text{sec}}\right)^2}{(2)\left(32.2\ \frac{\text{lbm-ft}}{\text{lbf-sec}^2}\right)}\\ &= 72.7\ \text{lbf}\end{aligned}$$

The additional power needed by the car traveling with the carrier is

$$P = F_D\mathrm{v} = \frac{(72.7\ \text{lbf})\left(110\ \frac{\text{ft}}{\text{sec}}\right)}{550\ \frac{\text{ft-lbf}}{\text{hp-sec}}} = 14.5\ \text{hp} \quad (15\ \text{hp})$$

The answer is (C).

15. $T_1 = 100°\text{F} + 460° = 560°\text{R}$

n is the polytropic exponent. For a polytropic process,

$$\begin{aligned}T_2 &= T_1\left(\frac{p_2}{p_1}\right)^{\frac{n-1}{n}} = (560°\text{R})\left(\frac{40\ \frac{\text{lbf}}{\text{in}^2}}{120\ \frac{\text{lbf}}{\text{in}^2}}\right)^{\frac{1.6-1}{1.6}}\\ &= 370.9°\text{R}\end{aligned}$$

The work done by the system per lbmol is given by

$$\begin{aligned}w &= \frac{R(T_1 - T_2)}{n-1} = \frac{\left(1545\ \frac{\text{ft-lbf}}{\text{lbmol-°R}}\right)(560°\text{R} - 370.9°\text{R})}{1.6-1}\\ &= 4.87\times 10^5\ \text{ft-lbf/lbmol}\end{aligned}$$

The molecular weight of air is 28.97 lbm/lbmol. The work per unit mass is

$$\begin{aligned}w &= \frac{4.87\times 10^5\ \frac{\text{ft-lbf}}{\text{lbmol}}}{28.97\ \frac{\text{lbm}}{\text{lbmol}}}\\ &= 1.68\times 10^4\ \text{ft-lbf/lbm} \quad (1.7\times 10^4\ \text{ft-lbf/lbm})\end{aligned}$$

The answer is (D).

16. Convert the resistivity to square inches. A circular mil is the area of a circle with a diameter of 0.001 in.

$$\begin{aligned}A_{\text{per cmil}} &= \frac{\pi}{4}d^2 = \left(\frac{\pi}{4}\right)(0.001\ \text{in})^2 = 7.854\times 10^{-7}\ \text{in}^2\\ \rho &= R_{\text{electrical}}A\\ &= \left(10\ \frac{\Omega\text{-cmil}}{\text{ft}}\right)\left(7.854\times 10^{-7}\ \frac{\text{in}^2}{\text{cmil}}\right)\\ &= 7.854\times 10^{-6}\ \Omega\text{-in}^2/\text{ft}\end{aligned}$$

The cross-sectional area of the copper tubing is

$$\begin{aligned}A_{\text{tube}} &= \frac{\pi}{4}(d_o^2 - d_i^2)\\ &= \frac{\pi}{4}\left((3.5\ \text{in})^2 - (3.062\ \text{in})^2\right)\\ &= 2.257\ \text{in}^2\end{aligned}$$

The maximum resistance that will lead to a drop of no more than 0.6 V is

$$R_{\max} = \frac{V_{\max}}{I} = \frac{0.6\ \text{V}}{35\ \text{A}} = 0.01714\ \Omega$$

The maximum length of the circuit is

$$L_{max} = \frac{R_{max}A_{tube}}{\rho} = \frac{(0.01714\ \Omega)(2.257\ \text{in}^2)}{7.854 \times 10^{-6}\ \frac{\Omega\text{-in}^2}{\text{ft}}} = 4926\ \text{ft}$$

Two wires (pipes) are needed to complete the circuit, so the maximum distance between power supply and pump is

$$d_{max} = \frac{L_{max}}{2} = \frac{4926\ \text{ft}}{2} = 2463\ \text{ft}\quad (2500\ \text{ft})$$

The answer is (B).

17. The power that must be supplied to the water is

$$P = \dot{V}\gamma h_p$$

Use the efficiencies of the motor and pump, and substitute known values.

$$P = \dot{V}\gamma h_p$$

$$(0.9)(0.8)P = \frac{\left(1500\ \frac{\text{gal}}{\text{min}}\right)\left(62.4\ \frac{\text{lbf}}{\text{ft}^3}\right)h_p}{\left(7.48\ \frac{\text{gal}}{\text{ft}^3}\right)\left(60\ \frac{\text{sec}}{\text{min}}\right)}$$

$$0.72P = \left(208.6\ \frac{\text{lbf}}{\text{sec}}\right)h_p$$

The pump head can be found using the Bernoulli equation.

$$\frac{p_1}{\gamma} + \frac{\text{v}_1^2}{2g} + z_1 + h_p = \frac{p_2}{\gamma} + \frac{\text{v}_2^2}{2g} + z_2 + h_L$$

Define points 1 and 2 to be the surfaces of the lower and upper tanks, respectively.

$$p_1 = p_2 = 0$$
$$\text{v}_1 = \text{v}_2 = 0$$

v can be found using the flow rate equation.

$$\dot{V} = \text{v}A$$

$$\frac{1500\ \frac{\text{gal}}{\text{min}}}{\left(7.48\ \frac{\text{gal}}{\text{ft}^3}\right)\left(60\ \frac{\text{sec}}{\text{min}}\right)} = \text{v}\left(\frac{\pi}{4}\right)\left(\frac{8\ \text{in}}{12\ \frac{\text{in}}{\text{ft}}}\right)^2$$

$$\text{v} = 9.57\ \text{ft/sec}$$

Substitute known values and the equation for the head loss in the pipe.

$$125\ \text{ft} + h_p = 200\ \text{ft} + 0.02\left(\frac{L}{D}\right)\left(\frac{\text{v}^2}{2g}\right)$$

$$h_p = 75\ \text{ft} + (0.02)\left(\frac{4000\ \text{ft}}{\frac{8\ \text{in}}{12\ \frac{\text{in}}{\text{ft}}}}\right)\left(\frac{\left(9.57\ \frac{\text{ft}}{\text{sec}}\right)^2}{(2)\left(32.2\ \frac{\text{ft}}{\text{sec}^2}\right)}\right) = 245.7\ \text{ft}$$

The power supplied to the motor is

$$0.72P = \frac{\left(208.6\ \frac{\text{lbf}}{\text{sec}}\right)(245.7\ \text{ft})\left(745.7\ \frac{\text{W}}{\text{hp}}\right)}{\left(550\ \frac{\text{ft-lbf}}{\text{hp-sec}}\right)\left(1000\ \frac{\text{W}}{\text{kW}}\right)}$$

$$P = 96.51\ \text{kW}\quad (97\ \text{kW})$$

The answer is (C).

18. Find the change in temperature. For water, $c \approx 1.0$ Btu/lbm-°F.

$$c = \frac{\Delta u}{\Delta T}$$
$$\Delta u = c\Delta T = \left(1.0\ \frac{\text{Btu}}{\text{lbm-°F}}\right)\Delta T$$

Multiply by the mass to find the total change in internal energy.

$$\Delta U = m\Delta u = mc\Delta T = (60\ \text{lbm})\left(1.0\ \frac{\text{Btu}}{\text{lbm-°F}}\right)\Delta T$$

To find ΔU, use the first law of thermodynamics.

$$Q - W = \Delta U$$

To achieve the maximum possible change in temperature, assume the tank is perfectly insulated so that $Q = 0$. W is negative because work is being done on the system.

$$-W = \Delta U$$

The total work, W, done by the mixing motor is

$$\begin{aligned} W &= Pt \\ &= (2\ \text{hp})\left(550\ \frac{\text{ft-lbf}}{\text{hp-sec}}\right)(15\ \text{min})\left(60\ \frac{\text{sec}}{\text{min}}\right) \\ &= 990{,}000\ \text{ft-lbf} \\ \Delta U &= \frac{990{,}000\ \text{ft-lbf}}{778\ \frac{\text{ft-lbf}}{\text{Btu}}} \\ &= 1272\ \text{Btu} \end{aligned}$$

Substitute to find ΔT.

$$\begin{aligned} \Delta U &= (60\ \text{lbm})\left(1.0\ \frac{\text{Btu}}{\text{lbm-°F}}\right)\Delta T \\ 1272\ \text{Btu} &= (60\ \text{lbm})\left(1.0\ \frac{\text{Btu}}{\text{lbm-°F}}\right)\Delta T \\ \Delta T &= 21.2\text{°F}\quad (21\text{°F}) \end{aligned}$$

The answer is (B).

19. Both the hole in the bearing and the shaft nose are dimensioned in decimals. The tolerance for decimal measurements is given in the illustration as 0.005 in. Under worst-case manufacturing conditions, the bearing hole would be 0.005 in deficient in diameter, or

$$b_{\text{bearing hole,worst}} = 0.600\ \text{in} - 0.005\ \text{in} = 0.595\ \text{in}$$

Under worst-case conditions, the shaft nose would be 0.005 in excessive in diameter, or

$$b_{\text{shaft nose,worst}} = 0.590\ \text{in} + 0.005\ \text{in} = 0.595\ \text{in}$$

The minimum clearance, then, is

$$\begin{aligned} C &= b_{\text{bearing hole,worst}} - b_{\text{shaft nose,worst}} \\ &= 0.595\ \text{in} - 0.595\ \text{in} \\ &= 0\ \text{in} \end{aligned}$$

A fit in which the shaft will always fit inside the hole is called a *clearance fit*. There is no interference between the parts, so this is not an *interference fit* (also known as a *press fit*). The term "tolerance fit" has no specific meaning in engineering.

The answer is (A).

20. Draw a diagram and a thermal circuit of the system.

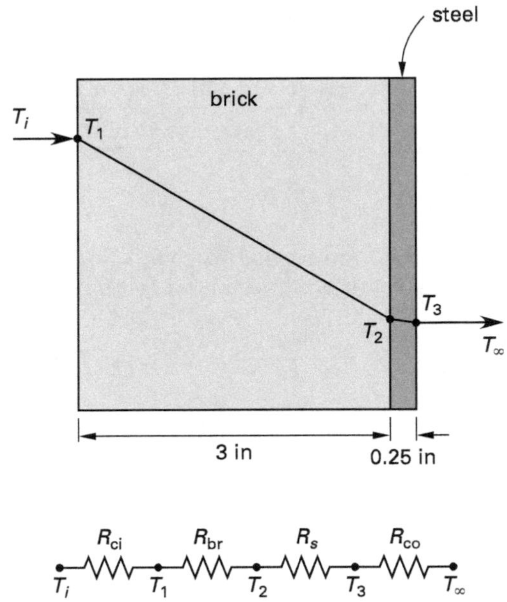

Use the given convection coefficient of $\bar{h}_{co} = 1.65$ Btu/hr-ft^2-°F. The thermal resistance, R_{co}, per square foot is

$$\begin{aligned} R_{co} &= \frac{1}{\bar{h}_{co}A} = \frac{1}{\left(1.65\ \frac{\text{Btu}}{\text{hr-ft}^2\text{-°F}}\right)(1\ \text{ft}^2)} \\ &= 0.606\ \text{hr-°F/Btu} \\ k_{br} &= 0.58\ \text{Btu-ft/hr-ft}^2\text{-°F} \\ k_s &= 26\ \text{Btu-ft/hr-ft}^2\text{-°F} \end{aligned}$$

(The value of k_s depends on the steel composition, but it does not affect the final result.)

$$\begin{aligned} R_{br} &= \frac{L_{br}}{Ak_{br}} = \frac{3\ \text{in}}{(1\ \text{ft}^2)\left(0.58\ \frac{\text{Btu-ft}}{\text{hr-ft}^2\text{-°F}}\right)\left(12\ \frac{\text{in}}{\text{ft}}\right)} \\ &= 0.431\ \text{hr-°F/Btu} \end{aligned}$$

$$\begin{aligned} R_s &= \frac{L_s}{Ak_s} = \frac{0.25\ \text{in}}{(1\ \text{ft}^2)\left(26\ \frac{\text{Btu-ft}}{\text{hr-ft}^2\text{-°F}}\right)\left(12\ \frac{\text{in}}{\text{ft}}\right)} \\ &= 0.0008\ \text{hr-°F/Btu} \\ T_1 &= 1000\text{°F} \\ T_\infty &= 70\text{°F} \end{aligned}$$

Since T_1 and T_∞ are the only known temperatures, the heat transfer is

$$\begin{aligned} q &= \frac{T_1 - T_\infty}{R_{\text{br}} + R_s + R_{\text{co}}} \\ &= \frac{1000°\text{F} - 70°\text{F}}{0.431\ \frac{\text{hr-°F}}{\text{Btu}} + 0.0008\ \frac{\text{hr-°F}}{\text{Btu}} + 0.606\ \frac{\text{hr-°F}}{\text{Btu}}} \\ &= 896\ \text{Btu/hr} \end{aligned}$$

The outside steel temperature is

$$\begin{aligned} q &= \frac{T_3 - T_\infty}{R_{\text{co}}} \\ 896\ \frac{\text{Btu}}{\text{hr}} &= \frac{T_3 - 70°\text{F}}{0.606\ \frac{\text{hr-°F}}{\text{Btu}}} \\ T_3 &= 613°\text{F} \quad (610°\text{F}) \end{aligned}$$

The answer is (C).

21. The temperature distribution in a single-pass, parallel-flow heat exchanger is shown.

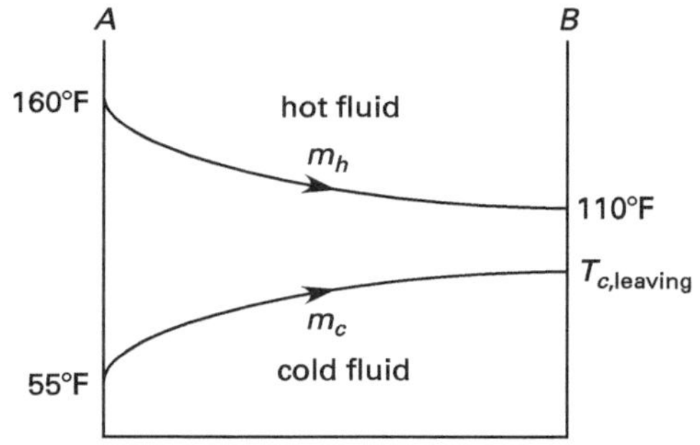

Find $T_{c,\text{leaving}}$.

$$\begin{aligned} q_h &= \dot{m}_h c_{p,h} \Delta T_h \\ &= \left(45{,}000\ \frac{\text{lbm}}{\text{hr}}\right)\left(0.9\ \frac{\text{Btu}}{\text{lbm-°F}}\right)(160°\text{F} - 110°\text{F}) \\ &= 2.025 \times 10^6\ \text{Btu/hr} \\ q_c &= \dot{m}_c c_{p,c} \Delta T_c \\ &= \left(40{,}000\ \frac{\text{lbm}}{\text{hr}}\right)\left(1\ \frac{\text{Btu}}{\text{lbm-°F}}\right)(T_{c,\text{leaving}} - 55°\text{F}) \\ &= \left(40{,}000\ \frac{\text{Btu}}{\text{hr-°F}}\right)(T_{c,\text{leaving}} - 55°\text{F}) \end{aligned}$$

The two heat flow rates are equal, so

$$\begin{aligned} q_h &= q_c \\ 2.025 \times 10^6\ \frac{\text{Btu}}{\text{hr}} &= \left(40{,}000\ \frac{\text{Btu}}{\text{hr-°F}}\right)(T_{c,\text{leaving}} - 55°\text{F}) \\ T_{c,\text{leaving}} &= 105.6°\text{F} \end{aligned}$$

The log mean temperature difference (LMTD) is

$$\begin{aligned} \text{LMTD} &= \frac{\Delta T_A - \Delta T_B}{\ln \frac{\Delta T_A}{\Delta T_B}} \\ &= \frac{(160°\text{F} - 55°\text{F}) - (110°\text{F} - 105.6°\text{F})}{\ln\left(\frac{160°\text{F} - 55°\text{F}}{110°\text{F} - 105.6°\text{F}}\right)} \\ &= 31.7°\text{F} \end{aligned}$$

The heat-transfer surface area of the heat exchanger is

$$\begin{aligned} q &= UA(\text{LMTD}) \\ 2.025 \times 10^6\ \frac{\text{Btu}}{\text{hr}} &= \left(75\ \frac{\text{Btu}}{\text{hr-ft}^2\text{-°F}}\right)A(31.7°\text{F}) \\ A &= 851.7\ \text{ft}^2 \quad (850\ \text{ft}^2) \end{aligned}$$

The answer is (B).

22. A pump curve represents the head a pump can supply for a given flow rate. To maintain a prescribed flow rate in a system, a certain amount of head must be supplied to the system. As the flow rate in the system increases, the amount of head that must be supplied also increases. By plotting the head needed by the system versus the flow rate (the system curve) on top of the pump curve, the point where the head produced by the pump is just the amount needed to overcome the head loss in the pipe can be found. The point where this occurs is where the two curves intersect and is known as the operating point.

An equation for the system curve can be obtained using the Bernoulli equation.

$$\frac{p_1}{\gamma} + \frac{\text{v}_1^2}{2g} + z_1 + h_p = \frac{p_2}{\gamma} + \frac{\text{v}_2^2}{2g} + z_2 + h_L$$

Define points 1 and 2 to be the surface of the lower and upper tanks, respectively.

$$\begin{aligned} p_1 &= p_2 = 0 \\ \text{v}_1 &= \text{v}_2 = 0 \end{aligned}$$

Substitute known values and the equation for the pipe head loss into the Bernoulli equation.

$$50 \text{ ft} + h_p = 150 \text{ ft} + \frac{fL\text{v}^2}{2Dg}$$

The inside diameter of 16 in schedule-40 pipe is 15 in.

The equivalent lengths of the fittings and valves are negligible compared to the total pipe length.

$$h_p = 100 \text{ ft} + \frac{(0.015)(3500 \text{ ft})\text{v}^2}{(2)\left(\frac{15 \text{ in}}{12 \frac{\text{in}}{\text{ft}}}\right)\left(32.2 \frac{\text{ft}}{\text{sec}^2}\right)}$$

Express v in terms of $\dot{V}$.

$$\dot{V} = \text{v}A = \text{v}\left(\frac{\pi}{4}\right)\left(\frac{15 \text{ in}}{12 \frac{\text{in}}{\text{ft}}}\right)^2$$

$$\text{v} = (0.815 \text{ ft}^{-2})\,\dot{V}$$

Substitute into the Bernoulli equation.

$$\begin{aligned} h_p &= 100 \text{ ft} + \left(\frac{(0.015)(3500 \text{ ft})\left((0.815 \text{ ft}^{-2})\,\dot{V}\right)^2}{(2)\left(\frac{15 \text{ in}}{12 \frac{\text{in}}{\text{ft}}}\right)\left(32.2 \frac{\text{ft}}{\text{sec}^2}\right)}\right) \\ &= 100 \text{ ft} + \left(0.433 \frac{\text{sec}^2}{\text{ft}^5}\right)\dot{V}^2 \end{aligned}$$

Plot the curve.

Q (gpm)	$\dot{V}$ (ft³/sec)	h_p (ft)
1200	2.674	103.1
2400	5.348	112.4
3600	8.022	127.9
4800	10.70	149.6
6000	13.37	177.4

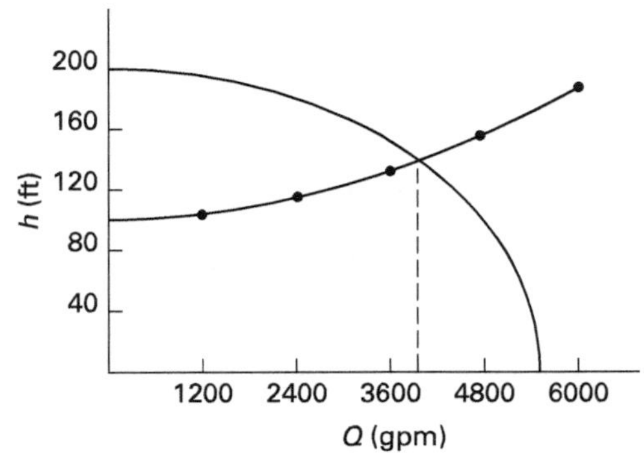

The discharge rate is approximately 4000 gpm.

$$Q = 4000 \text{ gpm}$$

The answer is (C).

23. Determine the fluid velocity through the pipe.

$$\begin{aligned} \text{v} &= \frac{\dot{V}}{A} = \frac{4\dot{V}}{\pi D^2} \\ &= \frac{(4)\left(0.185 \frac{\text{m}^3}{\text{min}}\right)}{\pi\left(\frac{19 \text{ mm}}{1000 \frac{\text{mm}}{\text{m}}}\right)^2\left(60 \frac{\text{s}}{\text{min}}\right)} \\ &= 10.9 \text{ m/s} \end{aligned}$$

The total minor loss is the sum of the loss contributions of each of the fittings.

$$
\begin{aligned}
h_{L,\text{minor}} &= \sum h_{\text{fitting}} = \sum \frac{\text{v}^2}{2g} k_{\text{fitting}} = \frac{\text{v}^2}{2g}(2k_{90°} + k_{45°}) \\
&= \left(\frac{\left(10.9 \ \frac{\text{m}}{\text{s}}\right)^2}{(2)\left(9.81 \ \frac{\text{m}}{\text{s}^2}\right)} \right)\big((2)(1.5) + 0.4\big) \\
&= 20.6 \text{ m} \quad (21 \text{ m})
\end{aligned}
$$

The answer is (C).

24. Define states 1 and 2′ as shown.

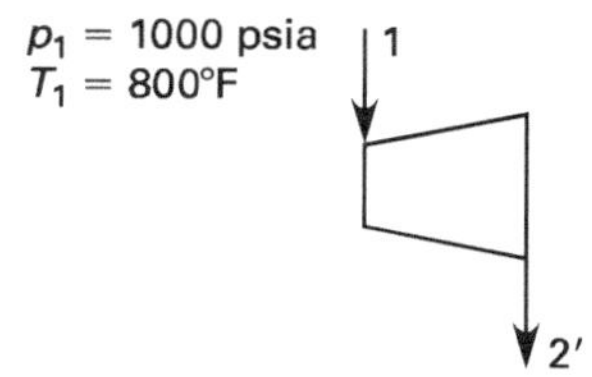

Using steam tables, $h_1 = 1388.5$ Btu/lbm, $s_1 = 1.5664$ Btu/lbm-°R, and $p_2 = 4$ psia. h_2 represents the enthalpy for a turbine that is 100% efficient. Since the turbine is isentropic, $s_1 = s_2$. Using the steam tables, find the appropriate enthalpy and entropy values at state 2′ where 2′ = 4 psia.

$$
\begin{aligned}
h_f &= 120.89 \text{ Btu/lbm} \\
s_f &= 0.21983 \text{ Btu/lbm-°R} \\
h_{fg} &= 1006.4 \text{ Btu/lbm} \\
s_{fg} &= 1.6426 \text{ Btu/lbm-°R}
\end{aligned}
$$

The steam quality at the turbine exhaust (state 2) for a 100% efficient turbine is found from the entropy relationship.

$$
\begin{aligned}
s &= s_f + x s_{fg} \\
1.5664 \ \frac{\text{Btu}}{\text{lbm-°R}} &= 0.21983 \ \frac{\text{Btu}}{\text{lbm-°R}} + x\left(1.6426 \ \frac{\text{Btu}}{\text{lbm-°R}}\right) \\
x &= 0.82
\end{aligned}
$$

The enthalpy at state 2, h_2, is

$$
\begin{aligned}
h_2 &= h_f + x h_{fg} = 120.89 \ \frac{\text{Btu}}{\text{lbm}} + (0.82)\left(1006.4 \ \frac{\text{Btu}}{\text{lbm}}\right) \\
&= 946.1 \text{ Btu/lbm}
\end{aligned}
$$

Since the turbine exhaust steam quality is 100%, the enthalpy at state 2′ is equal to the enthalpy of saturated vapor, h_g. From the steam tables at 4 psia,

$$h_2' = h_g = 1127.3 \text{ Btu/lbm}$$

The efficiency of the turbine is

$$
\begin{aligned}
\eta_{\text{turbine}} &= \frac{h_1 - h_2'}{h_1 - h_2} \\
&= \frac{1388.5 \ \frac{\text{Btu}}{\text{lbm}} - 1127.3 \ \frac{\text{Btu}}{\text{lbm}}}{1388.5 \ \frac{\text{Btu}}{\text{lbm}} - 946.1 \ \frac{\text{Btu}}{\text{lbm}}} \\
&= 0.59 \quad (59\%)
\end{aligned}
$$

Alternate Solution

The enthalpy h_2 can be found from a Mollier diagram, thereby avoiding two calculations. To do this, first find state 1 on the Mollier diagram. It will be at the intersection of the 1000 psia pressure curve and the 800°F curve. Since a 100% efficient turbine is an isentropic process, drop straight down on the diagram until the 4 psia pressure curve is crossed. This intersection represents state 2, and the enthalpy h_2 can be read as approximately 950 Btu/lbm.

The answer is (A).

25. The cross-sectional area of the cylinder is

$$
\begin{aligned}
A &= \frac{Q}{\text{v}} = \frac{\left(3 \ \frac{\text{gal}}{\text{min}}\right)\left(231 \ \frac{\text{in}^3}{\text{gal}}\right)}{\left(0.25 \ \frac{\text{ft}}{\text{sec}}\right)\left(12 \ \frac{\text{in}}{\text{ft}}\right)\left(60 \ \frac{\text{sec}}{\text{min}}\right)} \\
&= 3.85 \text{ in}^2
\end{aligned}
$$

Alternatively, use the standard (and not dimensionally consistent) formula.

$$
\begin{aligned}
\text{v}_{\text{ft/sec}} &= \frac{0.3208 Q_{\text{gpm}}}{A_{\text{in}^2}} \\
A_{\text{in}^2} &= \frac{0.3208 Q_{\text{gpm}}}{\text{v}_{\text{ft/sec}}} \\
&= \frac{(0.3208)\left(3 \ \frac{\text{gal}}{\text{min}}\right)}{0.25 \ \frac{\text{ft}}{\text{sec}}} \\
&= 3.85 \text{ in}^2
\end{aligned}
$$

Work is done by the cylinder as it compresses the soap cakes into bars. Work is equal to the force exerted by the cylinder multiplied by the distance it moves.

$$W = Fd$$

The force exerted by the cylinder is

$$F = \eta p A$$

The distance traveled by the cylinder is

$$d = \frac{W}{F} = \frac{W}{\eta p A} = \frac{(50\ \text{ft-lbf})\left(12\ \frac{\text{in}}{\text{ft}}\right)}{(0.85)\left(900\ \frac{\text{lbf}}{\text{in}^2}\right)(3.85\ \text{in}^2)}$$
$$= 0.2037\ \text{in} \quad (0.20\ \text{in})$$

The answer is (D).

26. Since the problem involves transient heat flow, determine if the lumped capacitance approximation is valid. The average temperature of the copper is

$$\overline{T} = \frac{T_s + T_i}{2} = \frac{450^\circ\text{C} + 200^\circ\text{C}}{2}$$
$$= 325^\circ\text{C}$$
$$T = 325^\circ\text{C} + 273^\circ = 598\text{K}$$

At this temperature, the sphere's conductivity, specific heat, and density are

$$k = 379\ \text{W/m·K}$$
$$c_p = 417\ \text{J/kg·K}$$
$$\rho = 8933\ \text{kg/m}^3$$

Calculate the sphere's characteristic length, L_c.

$$L_c = \frac{V}{A_s} = \frac{\frac{4}{3}\pi R^3}{4\pi R^2} = \frac{R}{3}$$
$$= \frac{100\ \text{mm}}{3}$$
$$= 33.3\ \text{mm} \quad (33.3 \times 10^{-3}\ \text{m})$$

Determine the Biot number, Bi.

$$\text{Bi} = \frac{hL_c}{k}$$
$$= \frac{\left(880\ \frac{\text{W}}{\text{m}^2\text{·K}}\right)(33.3 \times 10^{-3}\ \text{m})}{379\ \frac{\text{W}}{\text{m·K}}}$$
$$= 0.0773$$

Because the Biot number is less than 0.1, the internal thermal resistance of the sphere is negligible compared to the external thermal resistance in the oil bath. Therefore, the lumped parameter method can be used.

$$T_t = T_\infty + (T_0 - T_\infty)e^{-\text{BiFo}}$$

Solve for the Fourier number, Fo, and substitute its definition.

$$\text{Fo} = \frac{-1}{\text{Bi}}\ln\left(\frac{T_t - T_\infty}{T_0 - T_\infty}\right)$$
$$\frac{kt}{\rho c_p L_c^2} = \frac{-1}{\text{Bi}}\ln\left(\frac{T_t - T_\infty}{T_0 - T_\infty}\right)$$

Solve for t.

$$t = \frac{-\rho c_p L_c^2}{k\text{Bi}}\ln\left(\frac{T_t - T_\infty}{T_0 - T_\infty}\right)$$
$$= \frac{-\left(8933\ \frac{\text{kg}}{\text{m}^3}\right)\left(417\ \frac{\text{J}}{\text{kg·K}}\right)(3.33 \times 10^{-2}\ \text{m})^2}{\left(379\ \frac{\text{W}}{\text{m·K}}\right)(7.73 \times 10^{-2})}$$
$$\times \ln\left(\frac{200^\circ\text{C} - 75^\circ\text{C}}{450^\circ\text{C} - 75^\circ\text{C}}\right)$$
$$= 155\ \text{s} \quad (160\ \text{s})$$

The answer is (B).

27. The valve flow coefficient, C_v, relates flow rate to the pressure drop by the following equation.

$$Q_{\text{gpm}} = C_v\sqrt{\frac{\Delta p_{\text{psi}}}{\text{SG}}}$$

The valve flow coefficient is related to the loss coefficient by the equation

$$C_v = \frac{29.9 d_{\text{in}}^2}{\sqrt{K}}$$

Combining the two equations and solving for the valve diameter,

$$Q_{\text{gpm}} = \frac{29.9 d_{\text{in}}^2}{\sqrt{K}}\sqrt{\frac{\Delta p_{\text{psi}}}{\text{SG}}}$$
$$d_{\text{in}} = \sqrt{\frac{Q_{\text{gpm}}\sqrt{K}}{29.9\sqrt{\frac{\Delta p_{\text{psi}}}{\text{SG}}}}}$$

For a fully open gate valve, the loss coefficient is $K = 0.19$. The minimum needed diameter, then, is

$$d_{in} = \sqrt{\frac{Q_{gpm}\sqrt{K}}{29.9\sqrt{\frac{\Delta p_{psi}}{SG}}}} = \sqrt{\frac{\left(60\ \frac{gal}{min}\right)\sqrt{0.19}}{29.9\sqrt{\frac{5\ \frac{lbf}{in^2}}{1}}}} = 0.63\ in$$

Of the available choices, the smallest adequate choice is $d_{in} = 3/4$ in.

The answer is (B).

28. In order to lift a ball, the drag force created by the air on the ball needs to be greater than the ball's weight. The air velocity that makes the drag force equal to the weight will be the minimum velocity.

$$W = F_D = \frac{\rho C_D A v^2}{2g_c}$$

$$0.00516\ lbf = \frac{\left(0.075\ \frac{lbm}{ft^3}\right)C_D\left(\frac{\pi}{4}\right)\left(\frac{1.5\ in}{12\ \frac{in}{ft}}\right)^2 v^2}{(2)\left(32.2\ \frac{lbm\text{-}ft}{lbf\text{-}sec^2}\right)} = \left(1.43\times10^{-5}\ \frac{lbf\text{-}sec^2}{ft^2}\right)C_D v^2$$

The Reynolds number is needed to determine the drag coefficient, C_D, from a drag coefficient graph.

$$Re = \frac{vD}{\nu}$$

But to find the Reynolds number, the velocity, another unknown variable, is needed. Therefore, the solution is iterative and requires an assumed drag coefficient. On a graph of the Reynolds number for a sphere, the drag coefficient varies between 0.4 and 0.45 for Reynolds numbers between 1×10^3 and 1×10^5. Use an initial drag coefficient of 0.4. With this value, the minimum velocity becomes

$$0.00516\ lbf = \left(1.43\times10^{-5}\ \frac{lbf\text{-}sec^2}{ft^2}\right)(0.4)v^2$$

$$v = 30.0\ ft/sec$$

Check the Reynolds number to verify the assumed value of the drag coefficient.

$$Re = \frac{vD}{\nu} = \frac{\left(30.0\ \frac{ft}{sec}\right)\left(\frac{1.5\ in}{12\ \frac{in}{ft}}\right)}{1.58\times10^{-4}\ \frac{ft^2}{sec}} = 2.4\times10^4$$

With this Reynolds number, the drag coefficient for a sphere is approximately 0.42, close to the initial assumption.

The flow rate through the container is

$$\dot{V} = vA = \left(30.0\ \frac{ft}{sec}\right)(2\ ft)(2\ ft)\left(60\ \frac{sec}{min}\right) = 7200\ ft^3/min$$

The answer is (B).

29. The power generated by the turbine is

$$P = \dot{V}\gamma h_t$$

The velocity head of the jet is

$$h_t = \frac{v_j^2}{2g}$$

The turbine power becomes

$$P = \dot{V}\gamma\frac{v_j^2}{2g}$$

Find the flow rate.

$$\dot{V} = v_j A_j = v_j\left(\frac{\pi}{4}\right)(2\ ft)^2 = (\pi\ ft^2)v_j$$

Substitute known values.

$$\left(\frac{122\times 10^6\ \text{W}}{745.7\ \dfrac{\text{W}}{\text{hp}}}\right)\left(550\ \frac{\text{ft-lbf}}{\text{hp-sec}}\right) = (\pi\ \text{ft}^2)\text{v}_j\left(62.4\ \frac{\text{lbf}}{\text{ft}^3}\right)\left(\frac{\text{v}_j^2}{(2)\left(32.2\ \dfrac{\text{ft}}{\text{sec}^2}\right)}\right)$$

$$\text{v}_j = 309\ \text{ft/sec}$$

For ideal conditions, the tangential speed of the vanes is half the speed of the jet. Find the angular velocity of the turbine wheel and convert it to rpm.

$$\text{v}_{\text{vane}} = r\omega$$
$$0.5\text{v}_j = r\omega$$
$$(0.5)\left(309\ \frac{\text{ft}}{\text{sec}}\right) = (10\ \text{ft})\omega$$
$$\omega = 15.5\ \text{rad/sec}$$
$$= \left(\frac{1\ \text{rev}}{2\pi\ \text{rad}}\right)\left(15.5\ \frac{\text{rad}}{\text{sec}}\right)\left(60\ \frac{\text{sec}}{\text{min}}\right)$$
$$= 148\ \text{rpm}\quad (150\ \text{rpm})$$

The answer is (B).

30. The easiest way to locate the points of maximum and minimum pressure is to draw the hydraulic grade line (HGL).

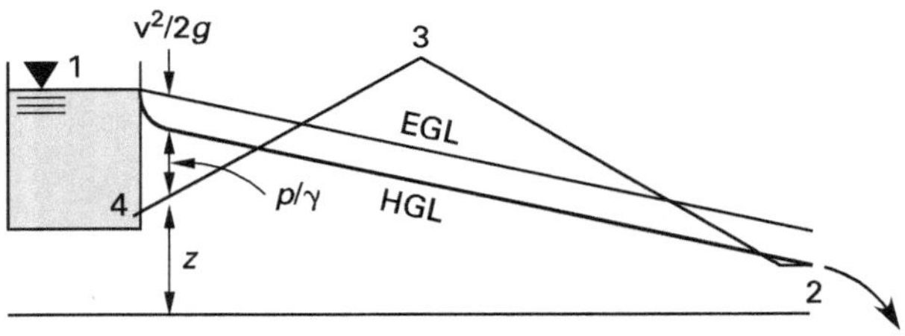

(not to scale)

The height p/γ represents the distance to which the liquid in a manometer would rise if a manometer were inserted into the pipe at that point. The height of the p/γ line along the length of the system is known as the hydraulic grade line, HGL. The total head in the system is greater than $p/\gamma + z$ by an amount $\text{v}^2/2g$, so the energy grade line, EGL, is above the HGL by a distance of $\text{v}^2/2g$. If the pipe rises above the HGL, then p/γ is negative. The advantage of drawing the HGL is that it makes apparent where the minimum and maximum pressures in the pipe occur.

The maximum pressure in the system occurs at point 4, before the water enters the pipe. Note the initial rapid drop in p/γ right after the water enters the pipe. This happens because some of the pressure head has been converted into velocity head. The maximum pressure in the system is simply the static pressure at point 4.

$$p_4 = \gamma h$$
$$= \left(62.4\ \frac{\text{lbf}}{\text{ft}^3}\right)\left(\frac{330\ \text{ft} - 315\ \text{ft}}{\left(12\ \dfrac{\text{in}}{\text{ft}}\right)^2}\right)$$
$$= 6.5\ \text{lbf/in}^2\quad (6.5\ \text{psi})$$

The minimum pressure will occur at point 3 since it is the furthest point above the HGL and will, therefore, be the most negative in the system. To compute the pressure at this point, however, the velocity of the water in the pipe must be calculated. The velocity can be found by writing the Bernoulli equation between points 1 and 2.

$$\frac{p_1}{\gamma} + \frac{\text{v}_1^2}{2g} + z_1 + h_p = \frac{p_2}{\gamma} + \frac{\text{v}_2^2}{2g} + z_2 + h_L$$
$$p_1 = p_2 = 0$$
$$\text{v}_1 = 0$$
$$h_p = 0$$

Substituting these values and the equation for h_L yields

$$330\ \text{ft} = \frac{\text{v}^2}{2g} + 100\ \text{ft} + 0.015\left(\frac{L}{D}\right)\left(\frac{\text{v}^2}{2g}\right)$$

$$230\ \text{ft} = \frac{\text{v}^2}{(2)\left(32.2\ \dfrac{\text{ft}}{\text{sec}^2}\right)} + (0.015)\left(\frac{1650\ \text{ft}}{2\ \text{ft}}\right) \times\left(\frac{\text{v}^2}{(2)\left(32.2\ \dfrac{\text{ft}}{\text{sec}^2}\right)}\right)$$

$$\text{v} = 33.3\ \text{ft/sec}$$

The minimum pressure can be found by writing the Bernoulli equation between points 1 and 3.

$$\frac{p_1}{\gamma} + \frac{\text{v}_1^2}{2g} + z_1 + h_p = \frac{p_3}{\gamma} + \frac{\text{v}_3^2}{2g} + z_3 + h_L$$
$$p_1 = 0$$
$$\text{v}_1 = 0$$
$$h_p = 0$$

Substituting these values and the equation for h_L yields

$$330\text{ ft} = \frac{p_3}{62.4\ \frac{\text{lbf}}{\text{ft}^3}} + \frac{\left(33.3\ \frac{\text{ft}}{\text{sec}}\right)^2}{(2)\left(32.2\ \frac{\text{ft}}{\text{sec}^2}\right)} + 333\text{ ft} + (0.015)\left(\frac{350\text{ ft}}{2\text{ ft}}\right)\left(\frac{\left(33.3\ \frac{\text{ft}}{\text{sec}}\right)^2}{(2)\left(32.2\ \frac{\text{ft}}{\text{sec}^2}\right)}\right)$$

$$p_3 = \frac{-4082\ \frac{\text{lbf}}{\text{ft}^2}}{\left(12\ \frac{\text{in}}{\text{ft}}\right)^2} = -28.3\text{ lbf/in}^2 \quad (-28\text{ psi})$$

The answer is (A).

31. Enthalpy is

$$h = u + \frac{pv}{J} = u + \frac{p}{J\rho}$$

The change in enthalpy from A to B is

$$\begin{aligned}\Delta h &= h_B - h_A = u_B - u_A + \frac{\frac{p_B}{\rho_B} - \frac{p_A}{\rho_A}}{J} \\ &= 53.1\ \frac{\text{Btu}}{\text{lbm}} - 119.1\ \frac{\text{Btu}}{\text{lbm}} \\ &\quad + \frac{\left(\frac{5.85\ \frac{\text{lbf}}{\text{in}^2}}{0.0508\ \frac{\text{lbm}}{\text{ft}^3}} - \frac{98.4\ \frac{\text{lbf}}{\text{in}^2}}{0.381\ \frac{\text{lbm}}{\text{ft}^3}}\right)\left(12\ \frac{\text{in}}{\text{ft}}\right)^2}{778\ \frac{\text{ft-lbf}}{\text{Btu}}} \\ &= -92.5\text{ Btu/lbm} \quad (-93\text{ Btu/lbm})\end{aligned}$$

The answer is (A).

32. Solve for the drag coefficient.

$$\begin{aligned}C_D &= \frac{4Dg(\rho_{\text{sphere}} - \rho_{\text{fluid}})}{3\text{v}^2\rho_{\text{fluid}}} \\ &= \frac{(4)(3\text{ in})\left(386.4\ \frac{\text{in}}{\text{sec}^2}\right)\left(488\ \frac{\text{lbm}}{\text{ft}^3} - 87\ \frac{\text{lbm}}{\text{ft}^3}\right)}{(3)\left(122\ \frac{\text{in}}{\text{sec}}\right)^2\left(87\ \frac{\text{lbm}}{\text{ft}^3}\right)} \\ &= 0.479 \quad (0.48)\end{aligned}$$

The answer is (D).

33. Use the Bernoulli equation.

$$\frac{p_1}{\gamma} + \frac{\text{v}_1^2}{2g} + z_1 + h_p = \frac{p_2}{\gamma} + \frac{\text{v}_2^2}{2g} + z_2 + h_L$$

Define point 1 as the surface of the tank and point 2 as the top of the fountain. At both points, the velocity of the stream is zero.

$$p_1 = p_2 = 0$$
$$\text{v}_1 = \text{v}_2 = 0$$

The Bernoulli equation simplifies to

$$z_1 = 4\text{ ft} + h_L$$

The head loss is

$$h_L = \frac{K_{\text{entering}}\text{v}^2}{2g} + \frac{fL\text{v}^2}{2gD}$$

The velocity of the water at the tank exit, and therefore in the pipe, is

$$\text{v} = \sqrt{2gh} = \sqrt{(2)\left(32.2\ \frac{\text{ft}}{\text{sec}^2}\right)(4\text{ ft})} = 16\text{ ft/sec}$$

For a sharp-edged exit,

$$K_{\text{entering}} = 0.5$$

The effective length is

$$\begin{aligned}L &= 40\text{ ft} + (5)(4.6\text{ ft}) + 2.5\text{ ft} \\ &= 65.5\text{ ft}\end{aligned}$$

Calculate the Reynolds number.

$$\text{Re}_D = \frac{\text{v}D}{\nu} = \frac{\left(16\ \frac{\text{ft}}{\text{sec}}\right)(0.3355\text{ ft})}{1.4\times 10^{-5}\ \frac{\text{ft}^2}{\text{sec}}} = 3.83\times 10^5$$

$$\frac{\epsilon}{D} = \frac{0.0002\text{ ft}}{0.3355\text{ ft}} = 0.0006$$

Using this Reynolds number and this relative roughness, the friction factor is found from the Moody chart to be $f = 0.019$. The head loss is

$$\begin{aligned} h_L &= \frac{K_{\text{entering}} \text{v}^2}{2g} + \frac{fL\text{v}^2}{2gD} \\ &= \frac{(0.5)\left(16 \ \frac{\text{ft}}{\text{sec}}\right)^2}{(2)\left(32.2 \ \frac{\text{ft}}{\text{sec}^2}\right)} + \frac{(0.019)(65.5 \text{ ft})\left(16 \ \frac{\text{ft}}{\text{sec}}\right)^2}{(2)\left(32.2 \ \frac{\text{ft}}{\text{sec}^2}\right)(0.3355 \text{ ft})} \\ &= 16.7 \text{ ft} \end{aligned}$$

The required depth in the tank is

$$\begin{aligned} z_1 &= 4 \text{ ft} + h_L = 4 \text{ ft} + 16.7 \text{ ft} \\ &= 20.7 \text{ ft} \quad (21 \text{ ft}) \end{aligned}$$

The answer is (B).

34. The volume of a fluid flowing through a pipe is the product of the velocity and the free cross-sectional area of the pipe.

$$\dot{V} = \text{v}A_{\text{pipe}}$$

The minimum cross-sectional area required to keep the water velocity under 7 ft/sec is

$$\begin{aligned} A_{\text{pipe}} &= \frac{\dot{V}}{\text{v}} = \frac{100 \ \frac{\text{gal}}{\text{min}}}{\left(7.0 \ \frac{\text{ft}}{\text{sec}}\right)\left(60 \ \frac{\text{sec}}{\text{min}}\right)\left(7.48 \ \frac{\text{gal}}{\text{ft}^3}\right)} \\ &= 0.0318 \text{ ft}^2 \end{aligned}$$

The smallest schedule-40 pipe that provides the required area is 2½ in nominal diameter, which has a cross-sectional area of 0.03325 ft^2.

The answer is (B).

35. Find the stroke length in feet.

$$L = \frac{8 \text{ in}}{12 \ \frac{\text{in}}{\text{ft}}} = 0.667 \text{ ft}$$

Find the bore area in square inches.

$$A = \frac{\pi}{4}D^2 = \left(\frac{\pi}{4}\right)(8 \text{ in})^2 = 50.27 \text{ in}^2$$

Find the number of engine power strokes per minute. The Otto engine operates on a four-stroke cycle.

$$\begin{aligned} N &= \frac{(\text{no. of strokes per revolution})n \times (\text{no. of cylinders})}{\text{no. of strokes per power stroke}} \\ &= \frac{\left(2 \ \frac{\text{strokes}}{\text{rev}}\right)\left(400 \ \frac{\text{rev}}{\text{min}}\right)(2)}{4 \ \frac{\text{strokes}}{\text{power stroke}}} \\ &= 400 \text{ power strokes/min} \end{aligned}$$

Use the PLAN formula to solve for the brake mean effective pressure.

$$\begin{aligned} \text{BHP} &= (\text{BMEP})(LAN) \\ \text{BMEP} &= \frac{\text{BHP}}{LAN} \\ &= \frac{(33 \text{ hp})\left(33{,}000 \ \frac{\text{ft-lbf}}{\text{hp-min}}\right)}{(0.667 \text{ ft})(50.27 \text{ in}^2)\left(400 \ \frac{\text{power strokes}}{\text{min}}\right)} \\ &= 81.2 \text{ lbf/in}^2 \quad (81 \text{ psig}) \end{aligned}$$

The answer is (D).

36. Find the flow through the valve. The minimum velocity of the cylinder is

$$\text{v} = \frac{L}{t} = \frac{24 \text{ in}}{5 \text{ sec}} = 4.8 \text{ in/sec}$$

The effective area of the cylinder is

$$A = \frac{\pi}{4}(d_{\text{bore}}^2 - d_{\text{rod}}^2) = \frac{\pi}{4}\left((3 \text{ in})^2 - (1 \text{ in})^2\right) = 6.28 \text{ in}^2$$

The flow rate is

$$Q = \text{v}A = \frac{\left(4.8 \ \frac{\text{in}}{\text{sec}}\right)(6.28 \text{ in}^2)\left(60 \ \frac{\text{sec}}{\text{min}}\right)}{231 \ \frac{\text{in}^3}{\text{gal}}} = 7.830 \text{ gpm}$$

Valve flow is related to the pressure drop by the valve flow coefficient.

$$Q_{\text{gpm}} = C_v\sqrt{\frac{\Delta p_{\text{psi}}}{\text{SG}}}$$

$$\Delta p_{\text{psi}} = \left(\frac{Q_{\text{gpm}}}{C_v}\right)^2 \text{SG} = \left(\frac{7.830\ \frac{\text{gal}}{\text{min}}}{2.9}\right)^2 (0.85)$$

$$= 6.2 \text{ psi}$$

The answer is (A).

37. The work for a heat pump can be found from

$$\text{COP} = \frac{\dot{Q}_H}{P}$$

COP is the coefficient of performance. The minimum cost will occur if the heat pump operates at ideal efficiency. For a Carnot heat pump, the COP is

$$\text{COP} = \frac{1}{1 - \frac{T_L}{T_H}}$$

T is in degrees Rankine. The heat pump power is

$$\begin{aligned} P &= \left(1 - \frac{T_L}{T_H}\right)\dot{Q}_H \\ &= \left(1 - \frac{30°\text{F} + 460°}{78°\text{F} + 460°}\right)\left(900{,}000\ \frac{\text{Btu}}{\text{day}}\right) \\ &= 80{,}300 \text{ Btu/day} \end{aligned}$$

$$\begin{aligned} \text{cost per month} &= \frac{\left(80{,}300\ \frac{\text{Btu}}{\text{day}}\right)\left(0.037\ \frac{\$}{\text{kW-hr}}\right)\left(30\ \frac{\text{days}}{\text{month}}\right)}{3412\ \frac{\text{Btu}}{\text{kW-hr}}} \\ &= \$26.12 \quad (\$26) \end{aligned}$$

The answer is (B).

38. Assume the thermal conductivity of the uranium fuel plate is constant. For a flat plate with an internal heat source, constant thermal conductivity, and uniform heat generation, the heat source generation rate, G, is

$$G = -k\frac{d^2T}{dx^2}$$

The thermal conductivity of uranium is 20 Btu/hr-ft-°F.

$$9 \times 10^7\ \frac{\text{Btu}}{\text{hr-ft}^3} = -\left(20\ \frac{\text{Btu}}{\text{hr-ft}^3\text{-°F}}\right)\frac{d^2T}{dx^2}$$

$$\frac{d^2T}{dx^2} = -4.5 \times 10^6 \text{ °F/ft}^2$$

Integrating twice,

$$\frac{dT}{dx} = \left(-4.5 \times 10^6\ \frac{\text{°F}}{\text{ft}^2}\right)x + C_1$$

$$T = \left(-2.25 \times 10^6\ \frac{\text{°F}}{\text{ft}^2}\right)x^2 + C_1x + C_2$$

Use the boundary conditions to find the constants of integration. At $x = 0$, $T = 500°\text{F}$, so $C_2 = 500°\text{F}$. At $x = 0.25$ in, $T = 750°\text{F}$.

$$\begin{aligned} 750°\text{F} &= \left(-2.25 \times 10^6\ \frac{\text{°F}}{\text{ft}^2}\right)\left(\frac{0.25 \text{ in}}{12\ \frac{\text{in}}{\text{ft}}}\right)^2 \\ &\quad + C_1\left(\frac{0.25 \text{ in}}{12\ \frac{\text{in}}{\text{ft}}}\right) + 500°\text{F} \\ C_1 &= 58{,}875 \text{ °F/ft} \end{aligned}$$

The equation for T is

$$T = \left(-2.25 \times 10^6\ \frac{\text{°F}}{\text{ft}^2}\right)x^2 + \left(58{,}875\ \frac{\text{°F}}{\text{ft}}\right)x + 500°\text{F}$$

The temperature is maximum when $dT/dx = 0$.

$$0 = -\left(4.5 \times 10^6\ \frac{\text{°F}}{\text{ft}^2}\right)x + 58{,}875\ \frac{\text{°F}}{\text{ft}}$$

$$x = 0.0131 \text{ ft}$$

$$\begin{aligned} T_{\text{max}} &= \left(-2.25 \times 10^6\ \frac{\text{°F}}{\text{ft}^2}\right)(0.0131 \text{ ft})^2 \\ &\quad + \left(58{,}875\ \frac{\text{°F}}{\text{ft}}\right)(0.0131 \text{ ft}) + 500°\text{F} \\ &= 885°\text{F} \quad (890°\text{F}) \end{aligned}$$

The answer is (B).

39. The efficiency of the cycle is calculated using

$$\eta = \frac{W}{Q_H}$$

W is the net work of the cycle, which is the work done by the turbine, W_t, less the work input into the pump, W_p.

$$W = W_t - W_p$$

Q_H is the heat input to the system. Defining point 1 as the turbine entrance, point 2 as the turbine exit, point 3 as the pump entrance, and point 4 as the pump exit, the efficiency can be written in terms of the enthalpies.

$$\eta = \frac{W_t - W_p}{Q_H} = \frac{(h_1 - h_2) - (h_4 - h_3)}{h_1 - h_4}$$

From steam tables, for 900°F at 1000 psia,

$$h_1 = 1448.1 \text{ Btu/lbm}$$

Since expansion in the turbine is an isentropic process, h_2 is most conveniently found using a Mollier diagram.

$$h_2 = 1075 \text{ Btu/lbm}$$

The pump work is calculated from

$$W_p = h_4 - h_3 = v_f(p_4 - p_3)$$

v_f is the specific volume of the saturated liquid entering the pump. From the saturated steam tables for 20 psia,

$$\begin{aligned} v_f &= 0.01683 \text{ ft}^3\text{/lbm} \\ h_3 &= 196.19 \text{ Btu/lbm} \end{aligned}$$

h_3 is needed to find h_4.

$$\begin{aligned} h_4 - h_3 &= v_f(p_4 - p_3) \\ &= \frac{\left(0.01683 \dfrac{\text{ft}^3}{\text{lbm}}\right) \times \left(1000 \dfrac{\text{lbf}}{\text{in}^2} - 20 \dfrac{\text{lbf}}{\text{in}^2}\right)\left(12 \dfrac{\text{in}}{\text{ft}}\right)^2}{778 \dfrac{\text{ft-lbf}}{\text{Btu}}} \\ &= 3.1 \text{ Btu/lbm} \end{aligned}$$

$$\begin{aligned} h_4 - 196.19 \frac{\text{Btu}}{\text{lbm}} &= 3.1 \text{ Btu/lbm} \\ h_4 &= 199.3 \text{ Btu/lbm} \end{aligned}$$

Substituting into the efficiency equation,

$$\begin{aligned} \eta &= \frac{(h_1 - h_2) - (h_4 - h_3)}{h_1 - h_4} \\ &= \frac{\left(1448.1 \dfrac{\text{Btu}}{\text{lbm}} - 1075 \dfrac{\text{Btu}}{\text{lbm}}\right) - 3.1 \dfrac{\text{Btu}}{\text{lbm}}}{1448.1 \dfrac{\text{Btu}}{\text{lbm}} - 199.3 \dfrac{\text{Btu}}{\text{lbm}}} \\ &= 0.296 \quad (30\%) \end{aligned}$$

The answer is (C).

40. Since two T_{out} values are unknown, this is an NTU (heat exchanger effectiveness) problem.

Assume $c_p = 1.0$ Btu/lbm-°F for water.

$$\begin{aligned} C_{\text{water}} &= \dot{m}c_p = \left(12{,}000 \frac{\text{lbm}}{\text{hr}}\right)\left(1.0 \frac{\text{Btu}}{\text{lbm-°F}}\right) \\ &= 12{,}000 \text{ Btu/hr-°F} \\ C_{\text{oil}} &= \dot{m}c_p = \left(20{,}000 \frac{\text{lbm}}{\text{hr}}\right)\left(0.60 \frac{\text{Btu}}{\text{lbm-°F}}\right) \\ &= 12{,}000 \text{ Btu/hr-°F} \end{aligned}$$

$$\frac{C_{\text{min}}}{C_{\text{max}}} = \frac{12{,}000 \dfrac{\text{Btu}}{\text{hr-°F}}}{12{,}000 \dfrac{\text{Btu}}{\text{hr-°F}}} = 1.0$$

$$\text{NTU} = \frac{UA}{C_{\text{min}}} = \frac{\left(280 \dfrac{\text{Btu}}{\text{hr-ft}^2\text{-°F}}\right)(40 \text{ ft}^2)}{12{,}000 \dfrac{\text{Btu}}{\text{hr-°F}}} = 0.93$$

From an NTU-effectiveness chart,

$$\begin{aligned} \epsilon &\approx 0.45 \\ \epsilon &= \frac{C_{\text{oil}}(T_{\text{oil,in}} - T_{\text{oil,out}})}{C_{\text{min}}(T_{\text{oil,in}} - T_{\text{cold,in}})} \\ 0.45 &= \frac{\left(12{,}000 \dfrac{\text{Btu}}{\text{hr-°F}}\right)(280\text{°F} - T_{\text{oil,out}})}{\left(12{,}000 \dfrac{\text{Btu}}{\text{hr-°F}}\right)(280\text{°F} - 60\text{°F})} \\ T_{\text{oil,out}} &= 181\text{°F} \quad (180\text{°F}) \end{aligned}$$

Since $C_{\min}/C_{\max} = 1.0$,

$$\epsilon = \frac{C_{\text{water}}(T_{\text{water,out}} - T_{\text{water,in}})}{C_{\min}(T_{\text{oil,in}} - T_{\text{water,in}})}$$

$$0.45 = \frac{\left(12{,}000\ \frac{\text{Btu}}{\text{hr-°F}}\right)(T_{\text{water,out}} - 60°\text{F})}{\left(12{,}000\ \frac{\text{Btu}}{\text{hr-°F}}\right)(280°\text{F} - 60°\text{F})}$$

$$T_{\text{water,out}} = 159°\text{F} \quad (160°\text{F})$$

Check.

$$\begin{aligned} q_{\text{water}} &= mc_p\Delta T \\ &= \left(12{,}000\ \frac{\text{lbm}}{\text{hr}}\right)\left(1.0\ \frac{\text{Btu}}{\text{lbm-°F}}\right)(159°\text{F} - 60°\text{F}) \\ &= 1.188 \times 10^6\ \text{Btu/hr} \end{aligned}$$

$$\begin{aligned} q_{\text{oil}} &= \left(20{,}000\ \frac{\text{lbm}}{\text{hr}}\right)\left(0.60\ \frac{\text{Btu}}{\text{lbm-°F}}\right)(280°\text{F} - 181°\text{F}) \\ &= 1.188 \times 10^6\ \text{Btu/hr} \quad [\text{OK}] \end{aligned}$$

The answer is (C).

41. The enthalpy of the output mixture is found from the conservation of energy equation.

$$\sum \dot{m}_{\text{entering}} h_{\text{entering}} = \sum \dot{m}_{\text{leaving}} h_{\text{leaving}}$$

The enthalpies of the two input steams are determined using steam tables.

$$\begin{aligned} p_1 &= 200\ \text{lbf/in}^2 \\ T_1 &= 600°\text{F} \\ h_1 &= 1322.3\ \text{Btu/lbm} \end{aligned}$$

For saturated liquid,

$$\begin{aligned} p_2 &= 200\ \text{lbf/in}^2 \\ h_f &= 355.5\ \text{Btu/lbm} \\ h_{fg} &= 843.3\ \text{Btu/lbm} \end{aligned}$$

The enthalpy of the second input steam is

$$\begin{aligned} h_2 &= h_f + xh_{fg} \\ &= 355.5\ \frac{\text{Btu}}{\text{lbm}} + (0.7)\left(843.3\ \frac{\text{Btu}}{\text{lbm}}\right) \\ &= 945.8\ \text{Btu/lbm} \end{aligned}$$

The enthalpy of the output mixture can now be found.

$$\sum \dot{m}_{\text{entering}} h_{\text{entering}} = \sum \dot{m}_{\text{leaving}} h_{\text{leaving}}$$

$$\left(4\ \frac{\text{lbm}}{\text{min}}\right)\left(1322.3\ \frac{\text{Btu}}{\text{lbm}}\right) + \left(11\ \frac{\text{lbm}}{\text{min}}\right)\left(945.8\ \frac{\text{Btu}}{\text{lbm}}\right) = \left(4\ \frac{\text{lbm}}{\text{min}} + 11\ \frac{\text{lbm}}{\text{min}}\right)h_{\text{leaving}}$$

$$h_{\text{leaving}} = 1046.2\ \text{Btu/lbm}$$

The output is partially saturated vapor at 200 psi. The quality of the output mixture, x, can be found using

$$\begin{aligned} h_{\text{leaving}} &= h_f + xh_{fg} \\ 1046.2\ \frac{\text{Btu}}{\text{lbm}} &= 355.5\ \frac{\text{Btu}}{\text{lbm}} + x\left(843.3\ \frac{\text{Btu}}{\text{lbm}}\right) \\ x &= 0.819 \quad (82\%) \end{aligned}$$

The answer is (B).

42. Draw a diagram and a thermal circuit of the system.

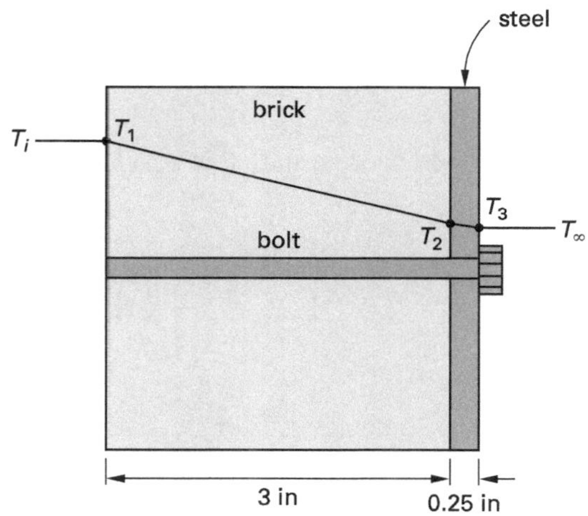

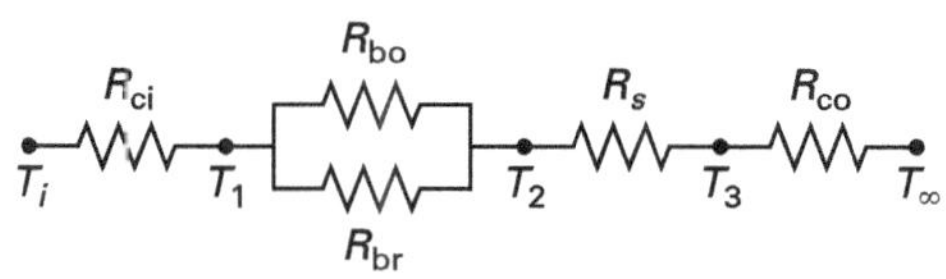

The heat flux can be found using

$$q = \frac{\Delta T}{\sum R} = \frac{T_1 - T_\infty}{R_{\text{eq}} + R_s + R_{\text{co}}}$$

T_1 and T_∞ are used since these are two known temperatures. The equivalent resistance of the parallel resistor combination of the bolt, R_{bo}, and the brick, R_{br}, is

$$R_{\text{eq}} = \frac{1}{\frac{1}{R_{\text{bo}}} + \frac{1}{R_{\text{br}}}}$$

Using $A = 1\ \text{ft}^2$ and a convection coefficient of $\bar{h}_{co} = 1.65\ \text{Btu/hr-ft}^2\text{-°F}$, R_{co} is

$$R_{co} = \frac{1}{\bar{h}_{co}A} = \frac{1}{\left(1.65\ \frac{\text{Btu}}{\text{hr-ft}^2\text{-°F}}\right)(1\ \text{ft}^2)} = 0.606\ \text{hr-°F/Btu}$$

The thermal conductivity of steel is

$$k_s = 26\ \text{Btu-ft/hr-ft}^2\text{-°F}$$

The thermal resistance of the steel per square foot is

$$R_s = \frac{L_s}{Ak_s} = \frac{0.25\ \text{in}}{(1\ \text{ft}^2)\left(26\ \frac{\text{Btu-ft}}{\text{hr-ft}^2\text{-°F}}\right)\left(12\ \frac{\text{in}}{\text{ft}}\right)} = 0.0008\ \text{hr-°F/Btu}$$

The thermal conductivity of fire-clay brick is

$$k_{br} = 0.58\ \text{Btu-ft/hr-ft}^2\text{-°F}$$

The area of the brick per square foot of furnace wall is

$$A_{br} = 1\ \text{ft}^2 - A_{bo} = 1\ \text{ft}^2 - (6)\left(\frac{\pi}{4}\right)\left(\frac{0.25\ \text{in}}{12\ \frac{\text{in}}{\text{ft}}}\right)^2 = 0.998\ \text{ft}^2$$

$$R_{br} = \frac{L}{Ak} = \frac{3\ \text{in}}{(0.998\ \text{ft}^2)\left(0.58\ \frac{\text{Btu-ft}}{\text{hr-ft}^2\text{-°F}}\right)\left(12\ \frac{\text{in}}{\text{ft}}\right)} = 0.432\ \text{hr-°F/Btu}$$

The thermal resistance of a bolt is

$$R_{bo} = \frac{L}{Ak} = \frac{3\ \text{in}}{(0.002\ \text{ft}^2)\left(26\ \frac{\text{Btu-ft}}{\text{hr-ft}^2\text{-°F}}\right)\left(12\ \frac{\text{in}}{\text{ft}}\right)} = 4.81\ \text{hr-°F/Btu}$$

The equivalent resistance of the parallel combination is

$$R_{eq} = \frac{1}{\frac{1}{R_{bo}} + \frac{1}{R_{br}}} = \frac{1}{\frac{1}{4.81\ \frac{\text{hr-°F}}{\text{Btu}}} + \frac{1}{0.432\ \frac{\text{hr-°F}}{\text{Btu}}}} = 0.396\ \text{hr-°F/Btu}$$

The heat flux can now be calculated.

$$q = \frac{T_1 - T_\infty}{R_{eq} + R_s + R_{co}} = \frac{1000\text{°F} - 70\text{°F}}{0.396\ \frac{\text{hr-°F}}{\text{Btu}} + 0.0008\ \frac{\text{hr-°F}}{\text{Btu}} + 0.606\ \frac{\text{hr-°F}}{\text{Btu}}} = 927\ \text{Btu/hr} \quad (930\ \text{Btu/hr})$$

The answer is (D).

43. Flow rate through the pump is determined from its displacement, D, and rotational speed, n. Because the pump and motor are direct coupled, the pump's rotational speed equals the motor's. The flow rate is

$$Q = Dn = \left(1.93\ \frac{\text{in}^3}{\text{rev}}\right)\left(1200\ \frac{\text{rev}}{\text{min}}\right) = 2316\ \text{in}^3\text{/min}$$

The required pump power is related to the pressure and the flow rate.

$$P = \frac{pQ}{\eta_{pump}} = \frac{\left(2500\ \frac{\text{lbf}}{\text{in}^2}\right)\left(2316\ \frac{\text{in}^3}{\text{min}}\right)}{(0.85)\left(33{,}000\ \frac{\text{ft-lbf}}{\text{hp-min}}\right)\left(12\ \frac{\text{in}}{\text{ft}}\right)} = 17.20\ \text{hp}$$

The current drawn by the motor is

$$I = \frac{P}{\sqrt{3}\,V\eta_{motor}(\text{pf})} = \frac{(17.20\ \text{hp})\left(746\ \frac{\text{W}}{\text{hp}}\right)}{\sqrt{3}\,(208\ \text{V})(0.90)(0.85)} = 46.56\ \text{A} \quad (50\ \text{A})$$

The answer is (B).

44. The amount of heat lost due to convection is

$$q = hA(T_s - T_\infty) = h(\pi d^2)(260\text{°F} - 80\text{°F}) = h(\pi)\left(\frac{2\ \text{in}}{12\ \frac{\text{in}}{\text{ft}}}\right)^2(260\text{°F} - 80\text{°F})$$

The convection coefficient can be found using

$$h = \frac{(\text{Nu})k}{d}$$

The Nusselt number of the sphere (based on diameter) is

$$\text{Nu} = 2 + 0.45(\text{GrPr})^{0.25}$$

The Grashof number for a sphere (based on diameter) is

$$\begin{aligned}\text{Gr} &= \left(\frac{g\beta\rho^2}{\mu^2}\right)d^3(T_s - T_\infty)\\ &= \left(\frac{g\beta\rho^2}{\mu^2}\right)\left(\frac{2\ \text{in}}{12\ \frac{\text{in}}{\text{ft}}}\right)^3(260°\text{F} - 80°\text{F})\end{aligned}$$

The properties of air at the mean temperature of 170°F are found using tables.

$$\begin{aligned}k &= 0.0168\ \text{Btu/hr-ft-°F}\\ \text{Pr} &= 0.72\\ \frac{g\beta\rho^2}{\mu^2} &= 1.123\times10^6\ 1/\text{ft}^3\text{-°F}\end{aligned}$$

The Grashof and Nusselt numbers are

$$\begin{aligned}\text{Gr} &= \left(\frac{g\beta\rho^2}{\mu^2}\right)d^3(T_s - T_\infty)\\ &= \left(1.123\times10^6\ \frac{1}{\text{ft}^3\text{-°F}}\right)\left(\frac{2\ \text{in}}{12\ \frac{\text{in}}{\text{ft}}}\right)^3(260°\text{F} - 80°\text{F})\\ &= 9.4\times10^5\end{aligned}$$

$$\begin{aligned}\text{Nu} &= 2 + 0.45(\text{GrPr})^{0.25}\\ &= 2 + (0.45)\big((9.4\times10^5)(0.72)\big)^{0.25}\\ &= 14.9\end{aligned}$$

The convection coefficient is

$$\begin{aligned}h &= \frac{(\text{Nu})k}{d} = \frac{(14.9)\left(0.0168\ \frac{\text{Btu}}{\text{hr-ft-°F}}\right)\left(12\ \frac{\text{in}}{\text{ft}}\right)}{2\ \text{in}}\\ &= 1.5\ \text{Btu/hr-ft}^2\text{-°F}\end{aligned}$$

The heat loss is

$$\begin{aligned}q &= hA(T_s - T_\infty)\\ &= \left(\frac{1.5\ \frac{\text{Btu}}{\text{hr-ft}^2\text{-°F}}}{3.412\ \frac{\text{Btu}}{\text{hr-W}}}\right)\left(\pi\left(\frac{2\ \text{in}}{12\ \frac{\text{in}}{\text{ft}}}\right)^2\right)(260°\text{F} - 80°\text{F})\\ &= 6.9\ \text{W}\end{aligned}$$

Alternate Solution

The heat loss can also be found following the procedure given in the *Mechanical Engineering Reference Manual.*

$$h = \frac{kC(\text{GrPr})^n}{L}$$

The characteristic length for a sphere is the radius. So the Grashof number becomes

$$\begin{aligned}\text{Gr} &= \left(\frac{g\beta\rho^2}{\mu^2}\right)d^3\left(T_s - T_\infty\right)\\ &= \left(1.123\times10^6\ \frac{1}{\text{ft}^3\text{-°F}}\right)\left(\frac{1\ \text{in}}{12\ \frac{\text{in}}{\text{ft}}}\right)^3(260°\text{F} - 80°\text{F})\\ &= 1.17\times10^5\end{aligned}$$

Since the product GrPr is between 10^3 and 10^9,

$$\begin{aligned}C &= 0.53\\ n &= 0.25\end{aligned}$$

The convection coefficient becomes

$$\begin{aligned}h &= \frac{kC(\text{GrPr})^n}{L}\\ &= \frac{\left(0.0168\ \frac{\text{Btu}}{\text{hr-ft-°F}}\right)(0.53)\times\big((1.17\times10^5)(0.72)\big)^{0.25}\left(12\ \frac{\text{in}}{\text{ft}}\right)}{1\ \text{in}}\\ &= 1.82\ \text{Btu/hr-ft}^2\text{-°F}\end{aligned}$$

This yields a heat loss of

$$\begin{aligned}q &= hA(T_s - T_\infty)\\ &= \left(\frac{1.82\ \frac{\text{Btu}}{\text{hr-ft}^2\text{-°F}}}{3.412\ \frac{\text{Btu}}{\text{hr-W}}}\right)\left(\pi\left(\frac{2\ \text{in}}{12\ \frac{\text{in}}{\text{ft}}}\right)^2\right)(260°\text{F} - 80°\text{F})\\ &= 8.4\ \text{W}\end{aligned}$$

The closest option is 6.9 W.

The answer is (B).

45. Calculate the hydraulic power.

$$P_{\text{hyd}} = h\dot{m}\left(\frac{g}{g_c}\right) = \left(\frac{(300\ \text{ft})\left(1.5\ \frac{\text{lbm}}{\text{sec}}\right)}{550\ \frac{\text{ft-lbf}}{\text{hp-sec}}}\right)\left(\frac{32.2\ \frac{\text{ft}}{\text{sec}}}{32.2\ \frac{\text{lbm-ft}}{\text{lbf-sec}^2}}\right) = 0.818\ \text{hp}$$

Calculate the motor power.

$$P_{\text{motor}} = \frac{P_{\text{hyd}}}{\eta_{\text{motor}}\eta_{\text{pump}}} = \frac{0.818\ \text{hp}}{(0.95)(0.65)} = 1.3\ \text{hp}$$

Calculate the motor current.

$$I_{\text{motor}} = \frac{P_{\text{motor}}}{V_{\text{motor}}} = \left(\frac{1.32\ \text{hp}}{240\ \text{V}}\right)\left(746\ \frac{\text{W}}{\text{hp}}\right) = 4.1\ \text{A}$$

The answer is (B).

46. This is a two-dimensional, steady-state conduction problem. If the surface temperature of the insulation can be determined, then the heat loss through the ground can be found.

For a first approximation, assume the pipe is at the same temperature as the glycol solution. Maximum heat loss will occur where the pipe enters the ground from the pond, where the temperature differential between the insulation surface and the ground is greatest.

For 1 in schedule-40 pipe, the outside diameter is $d_o = 1.315$ in. The outside diameter of the wrapped pipe is $d_{o,\text{wrapped}} = 1.315\ \text{in} + (2)(0.125\ \text{in}) = 1.565$ in.

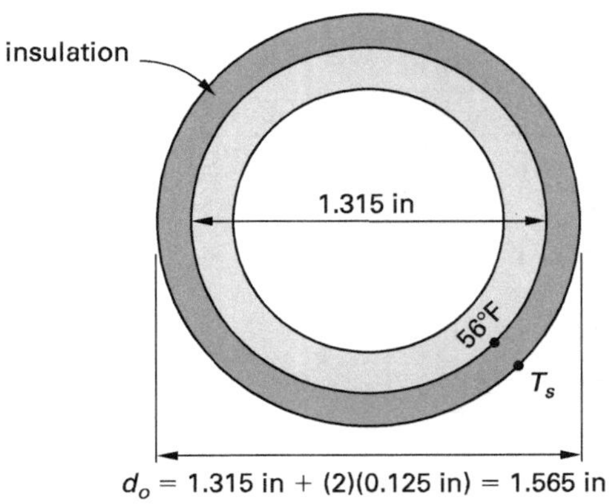

The resistance of the insulation per unit length can be found using

$$R = \frac{\ln\frac{r_o}{r_i}}{2\pi kL} = \frac{\ln\left(\frac{\frac{1.565\ \text{in}}{2}}{\frac{1.315\ \text{in}}{2}}\right)}{2\pi\left(0.03\ \frac{\text{Btu}}{\text{hr-ft-°F}}\right)L} = \frac{0.923\ \frac{\text{hr-ft-°F}}{\text{Btu}}}{L}$$

Let T_s be the temperature of the outside surface of the insulation. The heat transfer due to conduction through the insulation is

$$q = \frac{\Delta T}{R} = \frac{56°\text{F} - T_s}{\frac{0.923\ \frac{\text{hr-ft-°F}}{\text{Btu}}}{L}}$$

$$\frac{q}{L} = \left(1.08\ \frac{\text{Btu}}{\text{hr-ft-°F}}\right)(56°\text{F} - T_s)$$

The heat conduction through the ground is

$$q = Sk\Delta T$$

S is the shape factor. For a cylinder underground, S is

$$S = \frac{2\pi L}{\text{arccosh}\,\frac{2z}{d_o}} = \frac{2\pi L}{\text{arccosh}\left(\frac{(2)(3\ \text{ft})\left(12\ \frac{\text{in}}{\text{ft}}\right)}{1.565\ \text{in}}\right)} = 1.39L$$

Substitute known values into the heat conduction equation.

$$q = (1.39L)\left(1.5\ \frac{\text{Btu}}{\text{hr-ft-°F}}\right)(T_s - 20°\text{F})$$

$$\frac{q}{L} = \left(2.085\ \frac{\text{Btu}}{\text{hr-ft-°F}}\right)(T_s - 20°\text{F})$$

The heat transfer rates are equal.

$$\left(1.08\ \frac{\text{Btu}}{\text{hr-ft-°F}}\right)(56°\text{F} - T_s) = \left(2.085\ \frac{\text{Btu}}{\text{hr-ft-°F}}\right)(T_s - 20°\text{F})$$
$$T_s = 32.3°\text{F}$$

The heat transfer rate per unit length is

$$\frac{q}{L} = \left(2.085\ \frac{\text{Btu}}{\text{hr-ft-°F}}\right)(32.3°\text{F} - 20°\text{F}) = 25.6\ \text{Btu/hr-ft} \quad (26\ \text{Btu/hr-ft})$$

The answer is (D).

47. The rod will buckle when the thermal stress reaches the Euler stress. To find the Euler stress, check the slenderness ratio. The radius of gyration is

$$k = \sqrt{\frac{I}{A}} = \sqrt{\frac{\frac{\pi r^4}{4}}{\pi r^2}} = \frac{r}{2} = \frac{1.2\ \text{in}}{2} = 0.6\ \text{in}$$

For two built-in ends, the end restraint coefficient has a design value of $C = 0.65$. The slenderness ratio is

$$\frac{CL}{k} = \frac{(0.65)(11.5\ \text{ft})\left(12\ \frac{\text{in}}{\text{ft}}\right)}{0.6\ \text{in}} = 149.5$$

The rod may be regarded as a long column. The Euler stress for the rod is

$$\sigma_e = \frac{\pi^2 E}{\left(\frac{CL}{k}\right)^2}$$

The thermal stress for the rod can be expressed as a function of the temperature change.

$$\sigma_{\text{th}} = E\epsilon_{\text{th}} = E\alpha\Delta T$$

Find the temperature change that will cause the thermal stress to equal the Euler stress, by equating the thermal and Euler stresses and solving for the temperature change.

$$\sigma_{\text{th}} = \sigma_e$$
$$E\alpha\Delta T = \frac{\pi^2 E}{\left(\frac{CL}{k}\right)^2}$$
$$\Delta T = \frac{\pi^2}{\alpha\left(\frac{CL}{k}\right)^2}$$

The average coefficient of linear thermal expansion, α, for steel is $6.5 \times 10^{-6}\ 1/°\text{F}$.

$$\Delta T = \frac{\pi^2}{\alpha\left(\frac{CL}{k}\right)^2} = \frac{\pi^2}{\left(6.5 \times 10^{-6}\ \frac{1}{°\text{F}}\right)(149.5)^2} = 67.94°\text{F} \quad (68°\text{F})$$

The answer is (A).

48. First, construct a thermal circuit of the system.

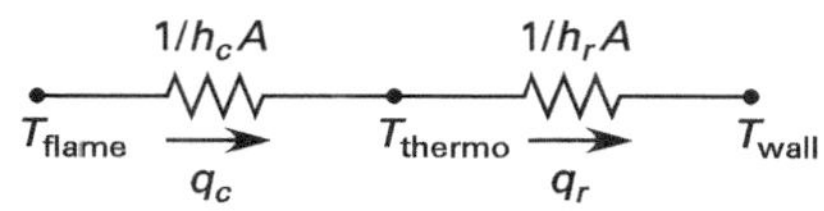

The heat transfer per unit area due to convection is

$$\frac{q_c}{A} = h_c(T_{\text{flame}} - T_{\text{thermo}}) = \left(25\ \frac{\text{Btu}}{\text{hr-ft}^2\text{-°F}}\right)(T_{\text{flame}} - 550°\text{F})$$

The heat transfer per unit area due to radiation is

$$\frac{q_r}{A} = \epsilon\sigma F_A(T^4_{\text{thermo}} - T^4_{\text{wall}})$$

The maximum flame temperature will be obtained by assuming all the surfaces to be black. The emissivity, ϵ, will then be equal to one. Since the thermocouple is in an enclosed surface with large area, the shape factor, F_A, is also equal to one.

$$T_{°R} = T_{°F} + 460°$$
$$T_{\text{flame}} = 550°F + 460° = 1010°R$$
$$T_{\text{wall}} = 400°F + 460° = 860°R$$
$$\frac{q_r}{A} = \epsilon\sigma F_A(T^4_{\text{thermo}} - T^4_{\text{wall}})$$
$$= (1)\left(0.1714 \times 10^{-8}\ \frac{\text{Btu}}{\text{hr-ft}^2\text{-°R}^4}\right)(1) \times \left((1010°R)^4 - (860°R)^4\right)$$
$$= 846\ \text{Btu/hr-ft}^2$$

The heat transfer rates are equal (i.e., $q_c = q_r$).

$$846\ \frac{\text{Btu}}{\text{hr-ft}^2} = \left(25\ \frac{\text{Btu}}{\text{hr-ft}^2\text{-°F}}\right)(T_{\text{flame}} - 550°F)$$
$$T_{\text{flame}} = 584°F \quad (580°F)$$

The answer is (A).

49. The heat transfer of a heat exchanger is directly proportional to the heat transfer coefficient, U. The heat transfer coefficient for a heat exchanger is in turn a direct function of the outside and inside convection coefficients, h_o and h_i, respectively, and the thermal conductivity, k.

$$\frac{1}{U} = f\left(\frac{1}{h_o} + \frac{1}{k} + \frac{1}{h_i}\right)$$

An increase in flow velocity results in an increase in the Reynolds number, and consequently an increase in the Nusselt number, which in turn results in an increase in h_o and h_i. Increasing h_o and h_i increases the value of U, increasing the heat transfer. Turbulent flow is desirable for heat transfer because it provides the highest possible convection coefficients.

Supersonic flow, though even faster, is not suitable for heat exchangers due to various risks such as cavitation.

The answer is (B).

50. The following diagrams apply to this problem.

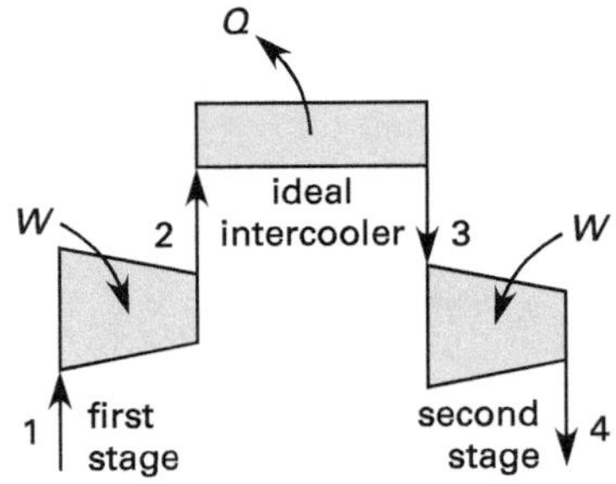

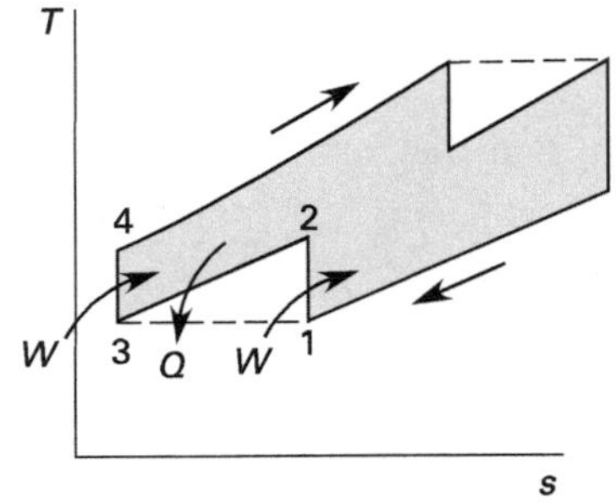

For an ideal intercooler,

$$T_1 = T_3 = 70°F$$
$$p_2 = p_3$$

The heat removed in the intercooler is

$$\dot{Q} = \dot{m}_a c_p(T_2 - T_3)$$
$$= \left(10\ \frac{\text{lbm}}{\text{min}}\right)\left(0.24\ \frac{\text{Btu}}{\text{lbm-°F}}\right)(T_2 - 70°F)$$

For isentropic compressions,

$$T_2 = T_1\left(\frac{p_2}{p_1}\right)^{\frac{k-1}{k}}$$

For air, $k = 1.4$, and T is in degrees Rankine.

$$T_{1,°R} = T_{°F} + 460° = 70°F + 460° = 530°R$$

$$T_2 = T_1\left(\frac{p_2}{p_1}\right)^{\frac{k-1}{k}} = (530°R)\left(\frac{p_2}{14.7\ \frac{\text{lbf}}{\text{in}^2}}\right)^{\frac{1.4-1}{1.4}}$$

Determine p_2. For ideal staging,

$$\frac{p_2}{p_1} = \frac{p_4}{p_3}$$
$$p_3 = p_2$$
$$p_2^2 = p_1 p_4$$
$$p_2 = \sqrt{p_1 p_4} = \sqrt{\left(14.7\ \frac{\text{lbf}}{\text{in}^2}\right)\left(200\ \frac{\text{lbf}}{\text{in}^2}\right)}$$
$$= 54.2\ \text{lbf/in}^2$$

Find T_2.

$$T_2 = (530°\text{R})\left(\frac{54.2\ \frac{\text{lbf}}{\text{in}^2}}{14.7\ \frac{\text{lbf}}{\text{in}^2}}\right)^{\frac{1.4-1}{1.4}} = 769°\text{R} - 460°$$
$$= 309°\text{F}$$

Now, $\dot{Q}$ can be found.

$$\dot{Q} = \dot{m}_a c_p (T_2 - T_3)$$
$$= \left(10\ \frac{\text{lbm}}{\text{min}}\right)\left(0.24\ \frac{\text{Btu}}{\text{lbm-°F}}\right)(309°\text{F} - 70°\text{F})$$
$$= 574\ \text{Btu/min} \quad (570\ \text{Btu/min})$$

The answer is (A).

51. Consider point 1 to be at the surface of the water in the tank and point 2 to be just outside the water's exit. Bernoulli's equation is

$$\frac{p_1}{\rho} + \frac{\text{v}_1^2}{2g_c} + \frac{z_1 g}{g_c} = \frac{p_2}{\rho} + \frac{\text{v}_2^2}{2g_c} + \frac{z_2 g}{g_c}$$

Both points are at atmospheric pressure, so the p/ρ terms are equal and can be eliminated. The surface of the water in the tank is stationary, so v_1 is zero, as is the term containing it. The remaining terms give

$$\frac{z_1 g}{g_c} = \frac{\text{v}_2^2}{2g_c} + \frac{z_2 g}{g_c}$$

Solving for the velocity at the exit,

$$\text{v}_2 = \sqrt{2g(z_1 - z_2)} = \sqrt{(2)\left(32.2\ \frac{\text{ft}}{\text{sec}^2}\right)(11\ \text{ft})}$$
$$= 26.62\ \text{ft/sec} \quad (27\ \text{ft/sec})$$

The answer is (C).

52. The following illustration shows the cooling-reheat process on a psychrometric chart. To remove moisture from the air, it is first cooled from state 1 to state 2′ at a constant humidity ratio, ω, so that the air becomes saturated ($\phi = 100\%$). At point 2′, the water vapor begins to condense, and further cooling to state 2 reduces the humidity ratio. The air is then heated to the desired temperature at the humidity ratio of state 2. This final point is state 3.

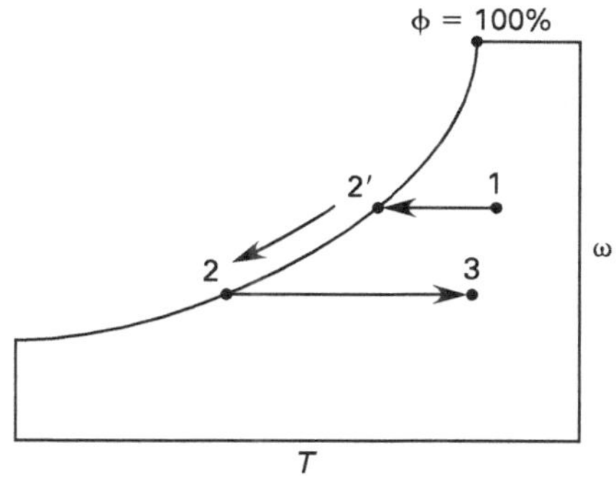

The heat removed during the cooling process is

$$\dot{q}_c = \dot{m}_a(h_1 - h_2 - (\omega_1 - \omega_2)h_f)$$

The term $(\omega_1 - \omega_2)h_f$ is small compared to other enthalpy terms and is often neglected.

Since there is no water lost or gained in the heating section of the evaporator, the heat added during the heating process is given by

$$\dot{q}_a = \dot{m}_a(h_3 - h_2)$$

The enthalpies h_1, h_2, and h_3 can be found using a psychrometric chart. First, locate point 1 where the dry-bulb temperature is 98°F, and then find the diagonal line sloping down from left to right for the 75°F wet-bulb temperature. Then, follow this diagonal line up and to the left to find $h_1 = 38.6$ Btu/lbm.

Second, locate point 3. This is the point where the dry-bulb temperature is 67°F and the relative humidity is 50%. Follow the diagonal line that passes through this point up and to the left to find $h_3 = 23.8$ Btu/lbm. To locate point 2, find point 3 again and then travel left on a horizontal line of constant specific humidity until the 100% humidity line is reached (the leftmost upward curving line). The intersection of the horizontal line and the 100% humidity line is point 2. Follow the diagonal line that passes through this point up and to the left to find $h_2 = 19.2$ Btu/lbm.

The air mass flow rate can be found from the volumetric flow rate and the specific volume, v, of the dry air at the state 1 (inlet) conditions. The specific volume of dry air can be found using the psychrometric chart.

$$\dot{m}_a = \frac{\dot{V}}{v_{a,1}} = \frac{800\ \frac{\text{ft}^3}{\text{min}}}{14.37\ \frac{\text{ft}^3}{\text{lbm}}}$$
$$= 55.7\ \text{lbm/min}$$

Substitute the air mass flow rate to find the rate at which heat is being removed.

$$q_r = \dot{m}_a(h_1 - h_2 - (\omega_1 - \omega_2)h_f)$$
$$= \left(55.7\ \frac{\text{lbm}}{\text{min}}\right)\left(60\ \frac{\text{min}}{\text{hr}}\right)\left(38.6\ \frac{\text{Btu}}{\text{lbm}} - 19.2\ \frac{\text{Btu}}{\text{lbm}}\right)$$
$$= 64{,}835\ \text{Btu/hr}\quad (65{,}000\ \text{Btu/hr})$$

Substitute the air mass flow rate to find the rate of heat addition.

$$q_a = \dot{m}_a(h_3 - h_2)$$
$$= \left(55.7\ \frac{\text{lbm}}{\text{min}}\right)\left(60\ \frac{\text{min}}{\text{hr}}\right)\left(23.8\ \frac{\text{Btu}}{\text{lbm}} - 19.2\ \frac{\text{Btu}}{\text{lbm}}\right)$$
$$= 15{,}373\ \text{Btu/hr}\quad (15{,}000\ \text{Btu/hr})$$

The answer is (D).

53. The following diagrams apply to this problem.

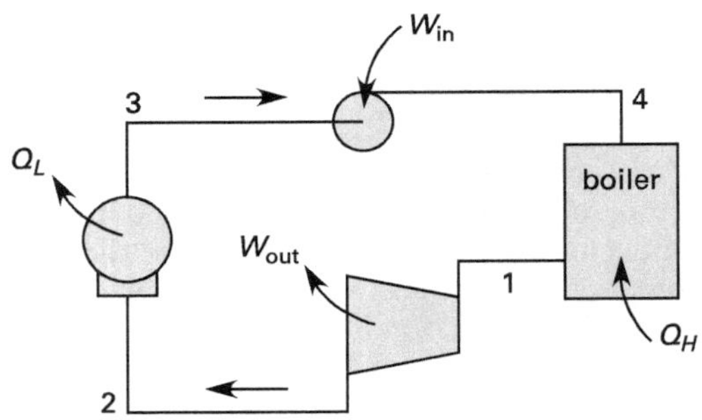

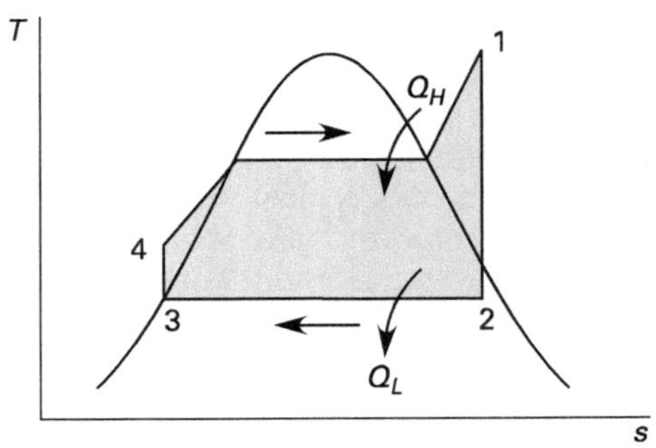

State 2 is at the output of the turbine where the temperature and steam quality are known. Since turbine expansion is an isentropic process, once the entropy at state 2 is found, the entropy at state 1 is also known. With this and the given pressure, the temperature of the steam entering the turbine can be found.

The entropy at state 2 can be found using

$$s_2 = s_f + xs_{fg}$$

Using the saturated steam tables at 100°F,

$$s_2 = s_f + xs_{fg}$$
$$= 0.12963\ \frac{\text{Btu}}{\text{lbm-°R}} + (0.92)\left(1.8526\ \frac{\text{Btu}}{\text{lbm-°R}}\right)$$
$$= 1.834\ \text{Btu/lbm-°R}$$

Since $s_1 = s_2$ and the pressure is 250 psia, using the superheated steam tables, the entering steam temperature is between 1040°F and 1060°F. This temperature can be found by linear interpolation between the 1040°F and 1060°F temperatures.

$$\frac{T_{1060°\text{F}} - T_{1040°\text{F}}}{s_{1060°\text{F}} - s_{1040°\text{F}}} = \frac{T_1 - T_{1040°\text{F}}}{s_1 - s_{1040°\text{F}}}$$

$$\frac{1060°\text{F} - 1040°\text{F}}{1.8386\ \frac{\text{Btu}}{\text{lbm-°R}} - 1.8315\ \frac{\text{Btu}}{\text{lbm-°R}}} = \frac{T_1 - 1040°\text{F}}{1.834\ \frac{\text{Btu}}{\text{lbm-°R}} - 1.8315\ \frac{\text{Btu}}{\text{lbm-°R}}}$$

$$T_1 = 1047°\text{F}\quad (1000°\text{F})$$

The answer is (C).

54. The T-s diagram for an air-standard Otto cycle is shown.

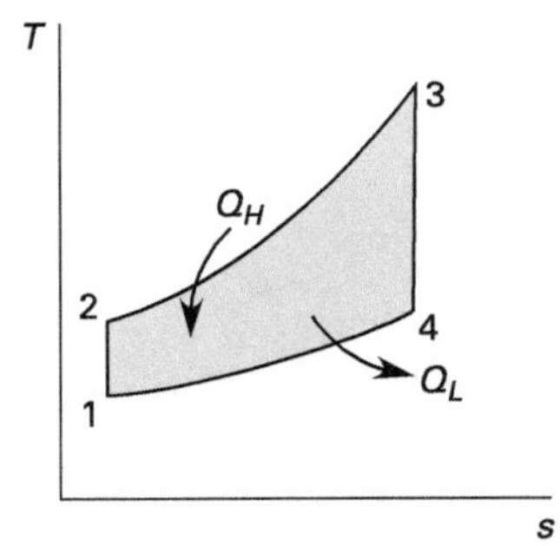

The thermal efficiency of an Otto cycle can be computed using

$$\eta_{\text{th}} = 1 - r^{1-k}$$

For air, $k = 1.4$.

$$\begin{aligned}\eta_{th} &= 1 - r^{1-k} \\ &= 1 - (8.5)^{1-1.4} \\ &= 0.575 \quad (57.5\%)\end{aligned}$$

The efficiency of a Carnot cycle is given by

$$T_{1,°R} = T_{°F} + 460° = 68°F + 460° = 528°R$$

$$\eta_{th} = 1 - \frac{T_L}{T_H} = 1 - \frac{528°R}{T_H}$$

T_H, the hottest temperature in the cycle, occurs at point 3, along with the highest pressure.

$$\frac{T_3}{T_2} = \frac{p_3}{p_2}$$

$$T_3 = T_2\left(\frac{510\ \frac{\text{lbf}}{\text{in}^2}}{p_2}\right)$$

The temperature T_2 is calculated using

$$T_2 = T_1\left(\frac{V_1}{V_2}\right)^{k-1}$$

V_1/V_2 is equal to the compression ratio.

$$T_2 = (528°R)(8.5)^{1.4-1} = 1243°R$$

Find the pressure p_2.

$$\begin{aligned}\frac{p_1V_1}{T_1} &= \frac{p_2V_2}{T_2} \\ p_2 &= \left(\frac{T_2p_1}{T_1}\right)\left(\frac{V_1}{V_2}\right) \\ &= \left(\frac{(1243°R)\left(14.7\ \frac{\text{lbf}}{\text{in}^2}\right)}{528°R}\right)(8.5) \\ &= 294\ \text{lbf/in}^2\end{aligned}$$

Substitute for p_2, and solve for T_3.

$$\begin{aligned}T_3 &= T_2\left(\frac{510\ \frac{\text{lbf}}{\text{in}^2}}{p_2}\right) = (1243°R)\left(\frac{510\ \frac{\text{lbf}}{\text{in}^2}}{294\ \frac{\text{lbf}}{\text{in}^2}}\right) \\ &= 2156°R\end{aligned}$$

The Carnot efficiency is

$$\begin{aligned}\eta_{th} &= 1 - \frac{528°R}{T_H} = 1 - \frac{528°R}{2156°R} \\ &= 0.755 \quad (75.5\%)\end{aligned}$$

The difference in efficiency is

$$\begin{aligned}\Delta\eta &= 75.5\% - 57.5\% \\ &= 18\% \quad (20\%)\end{aligned}$$

The answer is (B).

55. The following illustrations apply to this problem.

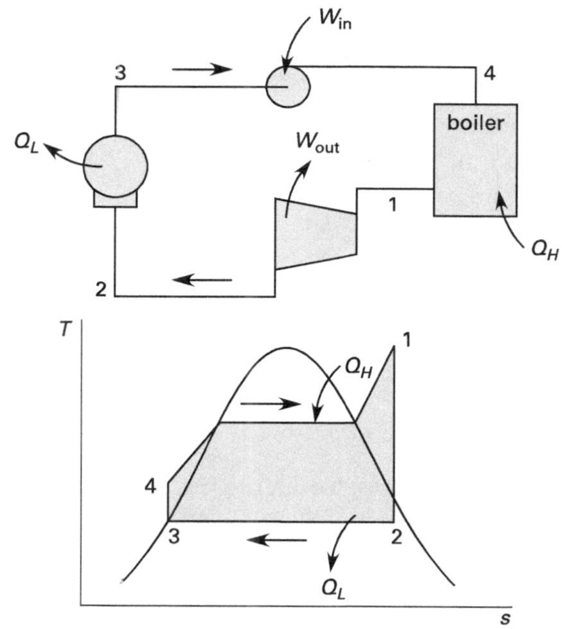

The mass flow rate can be determined using the net power output from the system.

$$\begin{aligned}P_{net} &= P_{turb} - P_{pump} \\ &= \dot{m}(h_1 - h_2) - \dot{m}(h_4 - h_3)\end{aligned}$$

The enthalpy at state 1 (1100°F and 1250 psia) is found in the superheated steam tables.

$$h_1 = 1556.6\ \text{Btu/lbm}$$

The enthalpy at state 2 is found using a Mollier diagram. Since the turbine is an isentropic process, locate state 1 on the diagram and drop straight down to the 5 psia curve. The enthalpy is

$$h_2 = 1018 \text{ Btu/lbm}$$

State 3 is saturated liquid at 5 psia.

$$h_3 = 130.17 \text{ Btu/lbm}$$

State 4 is compressed liquid. The enthalpy is

$$\begin{aligned} h_4 &= h_3 + v_3(p_4 - p_3) \\ &= 130.17\ \frac{\text{Btu}}{\text{lbm}} + \frac{\left(0.016407\ \frac{\text{ft}^3}{\text{lbm}}\right)\left(1250\ \frac{\text{lbf}}{\text{in}^2} - 5\ \frac{\text{lbf}}{\text{in}^2}\right) \times \left(12\ \frac{\text{in}}{\text{ft}}\right)^2}{778\ \frac{\text{ft-lbf}}{\text{Btu}}} \\ &= 133.95 \text{ Btu/lbm} \end{aligned}$$

Substitute and solve for the mass flow rate.

$$P_{\text{net}} = \dot{m}(h_1 - h_2) - \dot{m}(h_4 - h_3)$$

$$\begin{aligned} \left(250 \times 10^6 \text{ W}\right)\left(3.412\ \frac{\text{Btu}}{\text{hr-W}}\right) &= \dot{m}\left(1556.6\ \frac{\text{Btu}}{\text{lbm}} - 1018\ \frac{\text{Btu}}{\text{lbm}}\right) \\ &\quad - \dot{m}\left(133.95\ \frac{\text{Btu}}{\text{lbm}} - 130.17\ \frac{\text{Btu}}{\text{lbm}}\right) \\ \dot{m} &= 1.59 \times 10^6 \text{ lbm/hr} \quad (1.6 \times 10^6 \text{ lbm/hr}) \end{aligned}$$

Alternate Solution

Since the power output from the turbine is so large, the pump work into the system can be neglected.

$$\begin{aligned} (250 \times 10^6 \text{ W})\left(3.412\ \frac{\text{Btu}}{\text{hr-W}}\right) &= \dot{m}\left(1556.6\ \frac{\text{Btu}}{\text{lbm}} - 1018\ \frac{\text{Btu}}{\text{lbm}}\right) \\ \dot{m} &= 1.58 \times 10^6 \text{ lbm/hr} \quad (1.6 \times 10^6 \text{ lbm/hr}) \end{aligned}$$

The answer is (D).

56. The total number of spring coils includes both the active coils and the end coils. For squared and ground ends, the number of active coils is

$$n_{\text{active}} = n_{\text{total}} - 2$$

The force on the spring valve is a result of the pressure acting on the control area.

$$F = pA$$

The maximum spring deflection is

$$\delta = 7/16 \text{ in} = 0.4375 \text{ in}$$

The load deflection equation relates the properties of the spring to the applied force.

$$\begin{aligned} \frac{F}{\delta} &= \left(\frac{Gd}{8C^3}\right)\left(\frac{1}{n_{\text{active}}}\right) \\ n_{\text{active}} &= \frac{Gd\delta}{8C^3F} = \frac{Gd\delta}{8C^3(pA)} \\ &= \frac{\left(11.5 \times 10^6\ \frac{\text{lbf}}{\text{in}^2}\right)(0.10 \text{ in})(0.4375 \text{ in})}{(8)(8)^3\left(250\ \frac{\text{lbf}}{\text{in}^2}\right)(0.123 \text{ in}^2)} \\ &= 3.995 \\ n_{\text{total}} &= n_{\text{active}} + 2 = 3.995 + 2 \\ &= 5.995 \quad (6 \text{ coils}) \end{aligned}$$

The answer is (C).

57. The cycle and T-s diagrams are shown.

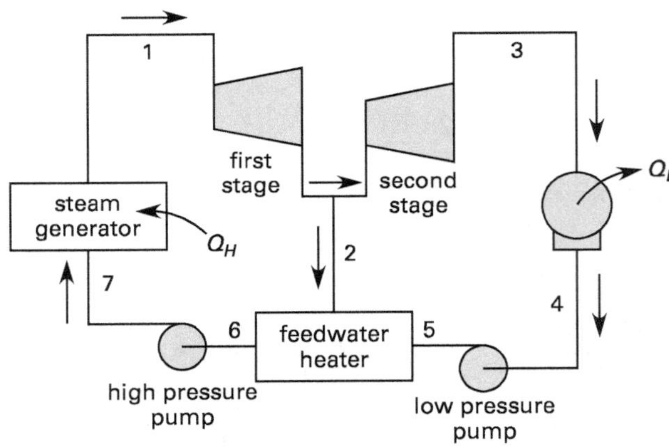

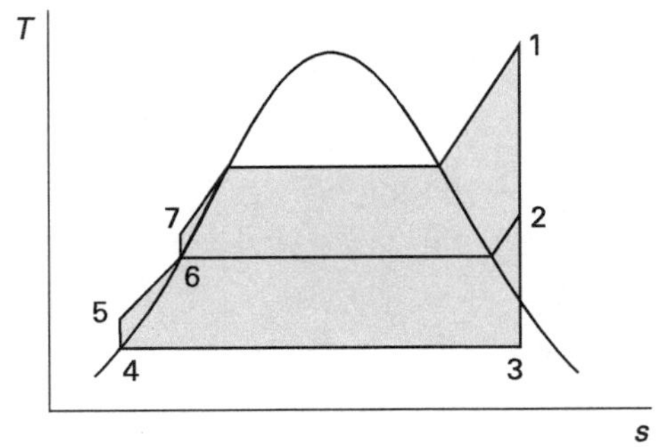

The fraction of the total flow passing through the second-stage turbine is

$$\frac{\dot{m}_3}{\dot{m}_1} = 1 - \frac{\dot{m}_2}{\dot{m}_1} = 1 - y$$

The fraction of total flow diverted to the open feedwater heater is y. The fraction y can be found from an energy balance for a control volume surrounding the feedwater heater.

$$y = \frac{h_6 - h_5}{h_2 - h_5}$$

Since $s_1 = s_2 = s_3$, the enthalpies at these states can be found from a Mollier diagram. h_1 and h_3 are not needed for this problem.

$$\begin{aligned} h_1 &= 1440 \text{ Btu/lbm} \quad [860°\text{F, 600 psia}] \\ h_2 &= 1260 \text{ Btu/lbm} \quad [130 \text{ psia}] \\ h_3 &= 1060 \text{ Btu/lbm} \quad [10 \text{ psia}] \end{aligned}$$

State 6 is saturated liquid. For an ideal open feedwater heater, $p_2 = p_5 = p_6 = 130$ psia. From the saturated liquid table,

$$h_6 = 319 \text{ Btu/lbm}$$

The enthalpy at state 4 is known since the water is saturated liquid at 10 psia. The enthalpy at state 5 can be found using the pump work.

$$h_5 - h_4 = v_f(p_5 - p_4)$$

From the saturated liquid table,

$$\begin{aligned} h_4 &= 161.23 \text{ Btu/lbm} \\ v_4 &= 0.01659 \text{ ft}^3\text{/lbm} \\ h_5 - h_4 &= v_f(p_5 - p_4) \end{aligned}$$

$$h_5 - 161.23\ \frac{\text{Btu}}{\text{lbm}} = \frac{\left(0.01659\ \frac{\text{ft}^3}{\text{lbm}}\right) \times \left(130\ \frac{\text{lbf}}{\text{in}^2} - 10\ \frac{\text{lbf}}{\text{in}^2}\right)\left(12\ \frac{\text{in}}{\text{ft}}\right)^2}{778\ \frac{\text{ft-lbf}}{\text{Btu}}}$$

$$h_5 = 161.6 \text{ Btu/lbm}$$

The fraction flowing through the second turbine is

$$\begin{aligned} \frac{\dot{m}_3}{\dot{m}_1} &= 1 - \frac{h_6 - h_5}{h_2 - h_5} \\ &= 1 - \frac{319\ \frac{\text{Btu}}{\text{lbm}} - 161.6\ \frac{\text{Btu}}{\text{lbm}}}{1260\ \frac{\text{Btu}}{\text{lbm}} - 161.6\ \frac{\text{Btu}}{\text{lbm}}} \\ &= 0.857 \quad (86\%) \end{aligned}$$

The answer is (D).

58. The specific heat for copper is 0.094 Btu/lbm-°F. The specific heat for stainless steel is 0.11 Btu/lbm-°F. The final temperature is

$$\begin{aligned} (mc_p\Delta T)_{\text{Cu}} &= (mc_p\Delta T)_{\text{st}} \\ (mc_p)_{\text{Cu}}(T_f - T_{i,\text{Cu}}) &= (mc_p)_{\text{st}}(T_{i,\text{st}} - T_f) \end{aligned}$$

$$\begin{aligned} (60 \text{ lbm})\left(0.094\ \frac{\text{Btu}}{\text{lbm-°F}}\right)(T_f - 50°\text{F}) \\ = (30 \text{ lbm})\left(0.11\ \frac{\text{Btu}}{\text{lbm-°F}}\right)(420°\text{F} - T_f) \\ T_f = 186.6°\text{F} \quad (190°\text{F}) \end{aligned}$$

The answer is (B).

59. The general combustion reaction for a hydrocarbon, C_nH_m, burning with theoretical air is

$$C_nH_m + xO_2 + 3.76xN_2 \rightarrow cCO_2 + dH_2O + 3.76xN_2$$

The solution is based on 1 kmol of fuel.

$$\begin{aligned} &0.8032CH_4 + 0.0575C_2H_6 + 0.0179C_3H_8 \\ &\quad +0.0169C_4H_{10} + 0.1045N_2 + xO_2 + 3.76xN_2 \\ &\quad \rightarrow c\begin{pmatrix} 0.079CO_2 + 0.001CO \\ +0.07O_2 + 0.85N_2 \end{pmatrix} + dH_2O \end{aligned}$$

The nitrogen ($3.76x$) has been grouped with the products.

Balance the carbon.

$$\begin{aligned} 0.8032 + (2)(0.0575) + (3)(0.0179) + (4)(0.0169) \\ = c(0.079 + 0.001) \\ c = 12.994 \end{aligned}$$

Balance the hydrogen.

$$\begin{aligned} (4)(0.8032) + (6)(0.0575) + (8)(0.0179) \\ +(10)(0.0169) \\ = 2d \\ d = 1.935 \end{aligned}$$

Balance the oxygen. (The variable x could also be found from a nitrogen balance.)

$$\begin{aligned} 2x &= (12.994)((2)(0.079) + 0.001 + (2)(0.07)) + 1.935 \\ x &= 2.91 \end{aligned}$$

The balanced equation is

$$\begin{aligned} &0.8032CH_4 + 0.0575C_2H_6 + 0.0179C_3H_8 \\ &\quad +0.01690C_4H_{10} + 0.1045N_2 + 2.91O_2 + 10.94N_2 \\ &\quad \rightarrow (12.994)\begin{pmatrix} 0.079CO_2 + 0.001CO \\ +0.07O_2 + 0.85N_2 \end{pmatrix} \\ &\quad +1.935H_2O \end{aligned}$$

The air/fuel ratio on a molar basis is

$$\begin{aligned} R_{\text{air/fuel}} &= \frac{n_{\text{air}}}{n_{\text{fuel}}} = \frac{n_{O_2} + n_{N_2}}{n_{\text{fuel}}} \\ &= \frac{2.91 \text{ kmol air} + 10.94 \text{ kmol air}}{1 \text{ kmol fuel}} \\ &= 13.85 \text{ kmol air/kmol fuel} \\ &\quad (14 \text{ kmol air/kmol fuel}) \end{aligned}$$

The answer is (D).

60. Convert the flow rate to cubic feet per second.

$$\dot{V} = \frac{1800 \ \frac{\text{gal}}{\text{min}}}{\left(7.48 \ \frac{\text{gal}}{\text{ft}^3}\right)\left(60 \ \frac{\text{sec}}{\text{min}}\right)} = 4.01 \text{ ft}^3\text{/sec}$$

Convert the diameter to feet.

$$d = \frac{10 \text{ in}}{12 \ \frac{\text{in}}{\text{ft}}} = 0.833 \text{ ft}$$

The velocity is

$$\begin{aligned} \text{v} &= \frac{\dot{V}}{A} = \frac{\dot{V}}{\pi\left(\frac{d}{2}\right)^2} = \frac{4.01 \ \frac{\text{ft}^3}{\text{sec}}}{\pi\left(\frac{0.833 \text{ ft}}{2}\right)^2} \\ &= 7.36 \text{ ft/sec} \end{aligned}$$

The head loss is

$$h_f = \frac{fL\text{v}^2}{2dg} = \frac{(0.0195)(185 \text{ ft})\left(7.36 \ \frac{\text{ft}}{\text{sec}}\right)^2}{(2)(0.833 \text{ ft})\left(32.2 \ \frac{\text{ft}}{\text{sec}^2}\right)} = 3.64 \text{ ft}$$

Convert the head loss to pressure drop.

$$\begin{aligned} \Delta p = h_f\rho\left(\frac{g}{g_c}\right) &= \frac{(3.64 \text{ ft})\left(62.4 \ \frac{\text{lbm}}{\text{ft}^3}\right)\left(\frac{32.2 \ \frac{\text{ft}}{\text{sec}^2}}{32.2 \ \frac{\text{lbm-ft}}{\text{lbf-sec}^2}}\right)}{\left(12 \ \frac{\text{in}}{\text{ft}}\right)^2} \\ &= 1.58 \text{ lbf/in}^2 \quad (1.6 \text{ psi}) \end{aligned}$$

The answer is (A).

61. The inner radius of the pipe is

$$\begin{aligned} r_1 &= \frac{d_1}{2} = \frac{4 \text{ in}}{2} \\ &= 2 \text{ in} \end{aligned}$$

The outer radius of the pipe is

$$\begin{aligned} r_2 &= \frac{d_2}{2} = \frac{4.5 \text{ in}}{2} \\ &= 2.25 \text{ in} \end{aligned}$$

The outer radius of the pipe with insulation is

$$\begin{aligned} r_3 &= \frac{d_3}{2} = \frac{4.5 \text{ in} + 1 \text{ in} + 1 \text{ in}}{2} \\ &= 3.25 \text{ in} \end{aligned}$$

Find the radial conduction through the pipe wall per foot of pipe.

$$\frac{Q}{L} = \frac{qA}{L} = \frac{2\pi(T_{\text{bulk}} - T_\infty)}{\frac{1}{r_a h_a} + \frac{\ln\frac{r_b}{r_a}}{k_{\text{pipe}}} + \frac{1}{r_b h_b} + \frac{\ln\frac{r_c}{r_b}}{k_{\text{insul}}} + \frac{1}{r_c h_c}}$$

$$\frac{Q}{L} = \left(\frac{2\pi(T_{\text{bulk}} - T_\infty)}{\dfrac{1}{r_a h_a} + \dfrac{\ln\dfrac{r_b}{r_a}}{k_{\text{pipe}}} + \dfrac{1}{r_b h_b} + \dfrac{\ln\dfrac{r_c}{r_b}}{k_{\text{insul}}} + \dfrac{1}{r_c h_c}} \right)$$

$$= \left(\frac{2\pi(300°\text{F} - 65°\text{F})}{\begin{array}{l} \dfrac{1}{(2.0\text{ in})(\infty)} + \dfrac{\ln\dfrac{2.25\text{ in}}{2.0\text{ in}}}{31\ \dfrac{\text{Btu-ft}}{\text{hr-ft}^2\text{-°F}}} \\ + \dfrac{1}{\left(\dfrac{2.25\text{ in}}{12\ \dfrac{\text{in}}{\text{ft}}}\right)\left(15\ \dfrac{\text{Btu}}{\text{hr-ft}^2\text{-°F}}\right)} \\ + \dfrac{\ln\dfrac{3.25\text{ in}}{2.25\text{ in}}}{0.03\ \dfrac{\text{Btu-ft}}{\text{hr-ft}^2\text{-°F}}} \\ + \dfrac{1}{\left(\dfrac{3.25\text{ in}}{12\ \dfrac{\text{in}}{\text{ft}}}\right)\left(2\ \dfrac{\text{Btu}}{\text{hr-ft}^2\text{-°F}}\right)} \end{array}} \right)$$

$$= 102\text{ Btu/hr-ft} \quad (100\text{ Btu/hr-ft})$$

The answer is (C).

62. The affinity laws indicate that when flow is reduced by 50%, speed is reduced by 50%, which in this case would be 1725 rpm. However, the pump is part of a larger application, and the system curve for this application does not start at zero head.

One-half the design flow is 500 gpm/2 = 250 gpm. According to the graph, 250 gpm occurs with a speed nearest to 2587 rpm (2600 rpm).

The answer is (C).

63. Location class 1 applies to any one-mile section containing 10 or fewer buildings intended for human occupancy. Typical areas in this class include wetlands, deserts, mountains, grazing land, farmland, and sparsely populated areas.

Location class 2 applies to any one-mile section containing more than 10, but fewer than 46, buildings intended for human occupancy. Typical areas in this class include fringe areas around cities and towns, industrial areas, ranches, and country estates.

Location class 3 applies to any one-mile section containing 46 or more buildings intended for human occupancy, except when location class 4 applies. Typical areas in this class include suburban housing developments, shopping centers, and residential areas.

Location class 4 applies to areas where most buildings are four stories and taller, where traffic is heavy or dense, and where a number of other utilities may be underground.

With 25 homes in the neighborhood, and none over two stories, location class 2 best applies to this area.

The answer is (B).

64. The PLAN formula can be used to determine the horsepower in an internal combustion engine operating on the Otto cycle.

$$P = pLAN$$

The number of power strokes per second is

$$N = \frac{2n(\text{no. cylinders})}{\text{no. strokes per cycle}}$$

$$= \frac{\left(2\ \dfrac{\text{strokes}}{\text{rev}}\right)\left(\dfrac{4200\ \dfrac{\text{rev}}{\text{min}}}{60\ \dfrac{\text{sec}}{\text{min}}}\right)(6\text{ cylinders})}{4\text{ strokes per power stroke}}$$

$$= 210\text{ power strokes/sec}$$

The power developed is

$$P = pLAN$$

$$= \frac{\begin{array}{l}\left(120\ \dfrac{\text{lbf}}{\text{in}^2}\right)\left(12\ \dfrac{\text{in}}{\text{ft}}\right)^2(84.4\text{ mm})\left(0.00328\ \dfrac{\text{ft}}{\text{mm}}\right) \\ \times\left(\dfrac{\pi}{4}\right)\left((100.4\text{ mm})\left(0.00328\ \dfrac{\text{ft}}{\text{mm}}\right)\right)^2 \\ \times\left(210\ \dfrac{\text{power strokes}}{\text{sec}}\right)\end{array}}{550\ \dfrac{\text{ft-lbf}}{\text{hp-sec}}}$$

$$= 155.6\text{ hp} \quad (160\text{ hp})$$

The answer is (B).

65. The only reason for pressure drop in this ductwork is friction. The pressure available to overcome friction will determine how fast the air will flow. For both the straight branch and the take-off branch, this pressure is equal to the pressure at the branch minus what is needed to drive the diffusers. This available pressure is

$$p_{\text{avail}} = 0.10\text{ in wg} - 0.03\text{ in wg} = 0.07\text{ in wg}$$

equal to the pressure at the branch minus what is needed to drive the diffusers. This available pressure is

$$\begin{aligned} p_{\text{avail}} &= 0.10 \text{ in wg} - 0.03 \text{ in wg} \\ &= 0.07 \text{ in wg} \end{aligned}$$

Convert the available pressure to inches of water gage per 100 ft of duct. For the 20 ft long straight branch,

$$\left(\frac{0.07 \text{ in wg}}{20 \text{ ft}}\right)(100 \text{ ft}) = 0.35 \text{ in wg}$$

From MERM Fig. "Standard Friction Loss in Standard Duct," using this pressure per 100 ft of duct and an air quantity of 1000 ft^3/min, the needed duct diameter is nearly 11 in.

For the 30 ft long take-off branch,

$$\left(\frac{0.07 \text{ in wg}}{30 \text{ ft}}\right)(100 \text{ ft}) = 0.23 \text{ in wg}$$

From the same MERM figure, using this pressure per 100 ft of duct and an air quantity of 1000 ft^3/min, the needed duct diameter is nearly 12 in.

The answer is (C).

66. The three-way valve shown is a mixing valve, which is commonly used to mix heating water to a desired temperature.

The answer is (C).

67. The energy lost can be determined from the energy balance for a reacting system.

$$\dot{Q}_{\text{CV}} = P_{\text{CV}} + \dot{n}_F(\bar{h}_P - \bar{h}_R)$$

The power generated is

$$P_{\text{CV}} = \frac{(110 \text{ hp})\left(2545 \frac{\text{Btu}}{\text{hp-hr}}\right)}{3600 \frac{\text{sec}}{\text{hr}}} = 77.76 \text{ Btu/sec}$$

The molar fuel flow rate is

$$\dot{n}_F = \frac{\dot{m}}{\text{MW}} = \frac{0.01 \frac{\text{lbm}}{\text{sec}}}{44 \frac{\text{lbm}}{\text{lbmol}}} = 2.27 \times 10^{-4} \text{ lbmol/sec}$$

The enthalpy of the reactants is simply the enthalpy of formation of the propane.

$$\bar{h}_R = (\bar{h}_f)_{\text{C}_3\text{H}_8} = -44{,}680 \text{ Btu/lbmol}$$

The enthalpy of the products is more complex and is

$$\begin{aligned} \bar{h}_P = {} & a(\bar{h}_f + \Delta\bar{h})_{\text{CO}_2} + b(\bar{h}_f + \Delta\bar{h})_{\text{H}_2\text{O}(g)} \\ & + (3.76x)(\bar{h}_f + \Delta\bar{h})_{\text{N}_2} \end{aligned}$$

The variables a, b, and x represent the quantity of products created. This must be determined by balancing the combustion equation. The general combustion reaction for a hydrocarbon, C_nH_m, burning with theoretical air is

$$\text{C}_n\text{H}_m + x\text{O}_2 + 3.76x\text{N}_2 \rightarrow a\text{CO}_2 + b\text{H}_2\text{O} + 3.76x\text{N}_2$$

The values for a, b, and x come from balancing the above equation and are

$$\begin{aligned} a &= n \\ 2b &= m \\ x &= n + \frac{b}{2} = n + \frac{m}{4} \end{aligned}$$

For propane, $n = 3$, and $m = 8$.

$$\begin{aligned} a &= 3 \\ b &= 4 \\ x &= 5 \end{aligned}$$

The balanced combustion equation is

$$\text{C}_3\text{H}_8 + 5\text{O}_2 + 18.8\text{N}_2 \rightarrow 3\text{CO}_2 + 4\text{H}_2\text{O} + 18.8\text{N}_2$$

The enthalpy of the products can be found.

$$\begin{aligned} T_{^\circ\text{R}} &= T_{^\circ\text{F}} + 460^\circ \\ T_{\text{air}} &= 77^\circ\text{F} + 460^\circ \\ &= 537^\circ\text{R} \\ T_{\text{air/fuel}} &= 1240^\circ\text{F} + 460^\circ \\ &= 1700^\circ\text{R} \end{aligned}$$

The enthalpy of the products is

$$\begin{aligned}\bar{h}_P &= a(\bar{h}_f + \Delta\bar{h})_{CO_2} + b(\bar{h}_f + \Delta\bar{h})_{H_2O(g)} \\ &\quad + (3.76x)(\bar{h}_f + \Delta\bar{h})_{N_2} \\ &= (3)\left(-169{,}300\ \frac{\text{Btu}}{\text{lbmol}} + \left(17{,}101.4\ \frac{\text{Btu}}{\text{lbmol}} - 4030.2\ \frac{\text{Btu}}{\text{lbmol}}\right)\right) \\ &\quad + (4)\left(-104{,}040\ \frac{\text{Btu}}{\text{lbmol}} + \left(14{,}455.4\ \frac{\text{Btu}}{\text{lbmol}} - 4258.3\ \frac{\text{Btu}}{\text{lbmol}}\right)\right) \\ &\quad + (18.8)\left(0\ \frac{\text{Btu}}{\text{lbmol}} + \left(12{,}178.9\ \frac{\text{Btu}}{\text{lbmol}} - 3729.5\ \frac{\text{Btu}}{\text{lbmol}}\right)\right) \\ &= -685{,}209\ \text{Btu/lbmol}\end{aligned}$$

The enthalpies of formation for the oxygen and nitrogen terms and the $\Delta\bar{h}$ terms for each of the reactants are zero because the air and fuel enter at 77°F.

The energy loss can now be found.

$$\begin{aligned}\dot{Q}_{CV} &= P_{CV} + \dot{n}_F(\bar{h}_P - \bar{h}_R) \\ &= 77.6\ \frac{\text{Btu}}{\text{sec}} + \left(2.27 \times 10^{-4}\ \frac{\text{lbmol}}{\text{sec}}\right) \\ &\quad \times \left(-685{,}209\ \frac{\text{Btu}}{\text{lbmol}} - \left(-44{,}680\ \frac{\text{Btu}}{\text{lbmol}}\right)\right) \\ &= -67.8\ \text{Btu/sec} \quad (-68\ \text{Btu/sec})\end{aligned}$$

The answer is (C).

68. The saturation temperature of the water vapor pressure is the dew point of the products. The water vapor pressure is a fraction of the total pressure of the products.

$$p_{water} = x_{water}p_{total} = x_{water}\left(14.7\ \frac{\text{lbf}}{\text{in}^2}\right)$$

The fraction of water vapor present can be obtained from the combustion equation. The general combustion reaction for a hydrocarbon, C_nH_m, burning with theoretical air is

$$C_nH_m + xO_2 + 3.76xN_2 \rightarrow aCO_2 + bH_2O + 3.76xN_2$$

In this case,

$$\begin{aligned} a &= n \quad \text{[from carbon balance]} \\ 2b &= m \quad \text{[from hydrogen balance]} \\ x &= n + \frac{m}{4} \quad \text{[from oxygen balance]}\end{aligned}$$

The composition of octane, C_8H_{18}, is $n = 8$ and $m = 18$, so

$$\begin{aligned} a &= 8 \\ b &= 9 \\ x &= 12.5\end{aligned}$$

The combustion reaction, then, for theoretical air is

$$\begin{aligned}&C_8H_{18} + 12.5O_2 + (3.76)(12.5)N_2 \\ &\quad \rightarrow 8CO_2 + 9H_2O + (3.76)(12.5)N_2\end{aligned}$$

The total number of moles of product is

$$\begin{aligned}N &= 8\ \frac{\text{mol}}{\text{mol-fuel}} + 9\ \frac{\text{mol}}{\text{mol-fuel}} + 47\ \frac{\text{mol}}{\text{mol-fuel}} \\ &= 64\ \text{mol/mol-fuel}\end{aligned}$$

The mole fraction of water is

$$x_{water} = \frac{b}{N} = \frac{9\ \frac{\text{mol}}{\text{mol-fuel}}}{64\ \frac{\text{mol}}{\text{mol-fuel}}} = 0.141$$

The water vapor pressure is

$$\begin{aligned}p_{water} &= x_{water}\left(14.7\ \frac{\text{lbf}}{\text{in}^2}\right) = (0.141)\left(14.7\ \frac{\text{lbf}}{\text{in}^2}\right) \\ &= 2.07\ \text{lbf/in}^2\end{aligned}$$

From the steam tables, the saturation temperature corresponding to this pressure is approximately

$$T_{dew} = 127°\text{F} \quad (130°\text{F}) \quad \text{[interpolated]}$$

The answer is (A).

69. The rate of conduction heat transfer is dependent on the difference in temperature between the two airstreams. The greater the temperature difference, the greater the rate of heat transfer.

As the airstreams transfer their heat, the warm airstream gets cooler and the cold airstream gets warmer. The closer the airstream is to the exit, then, the closer its temperature is to the other airstream. In quadrant I, the temperatures of the two airstreams are the closest, so the rate of heat transfer is the lowest.

The answer is (A).

70. To calculate the total present cost of the project, convert annual costs to equivalent present costs. The easiest method is to use tabulated factors, such as those in MERM App. "Cash Flow Equivalent Factors." To use the tabulated factors, however, annual costs must be uniform and must begin immediately. The operating costs can be expressed as the sum of two such annual costs, one of \$25,000 for five years, and one of −\$5000 for two years.

$$\begin{aligned} P_{\text{total cost}} &= P_{\text{initial}} + A_{\$25,000}(P/A, 15\%, 5) \\ &\quad + A_{-\$5000}(P/A, 15\%, 2) \\ &= \$100{,}000 + (\$25{,}000)(3.3522) \\ &\quad + (-\$5000)(1.6257) \\ &= \$175{,}677 \end{aligned}$$

Convert this present total cost to an annual amount, using an interest rate of 15%. This figure is the annual profit needed to achieve a return on investment of 15%.

$$\begin{aligned} A_{\text{total cost}} &= P_{\text{total cost}}(A/P, 15\%, 5) \\ &= (\$175{,}677)(0.2983) \\ &= \$52{,}404 \quad (\$52{,}000) \end{aligned}$$

The answer is (D).

71. When a fan is connected to an industrial ventilation system, the duct should be straight with no transitions for a certain distance in order to stabilize the airflow and achieve a uniform velocity profile. This distance, called the *100% effective duct length*, is calculated as

$$L_e = \frac{\sqrt{A}}{4.3} \quad [\text{if v} < 2500 \text{ ft/min}]$$

$$L_e = \frac{\text{v}\sqrt{A}}{10{,}600} \quad [\text{if v} > 2500 \text{ ft/min}]$$

In these equations, L_e is in feet, A is the cross-sectional area of the duct in square inches, and v is the airflow velocity in feet per minute. The equations are empirically derived, and the second is not dimensionally consistent.

To get the flow velocity, divide the volumetric flow rate by the duct area.

$$\begin{aligned} \text{v} &= \frac{\dot{V}}{A} = \frac{\dot{V}}{bh} \\ &= \frac{30{,}000 \ \frac{\text{ft}^3}{\text{min}}}{\left(\frac{36 \text{ in}}{12 \ \frac{\text{in}}{\text{ft}}}\right)\left(\frac{24 \text{ in}}{12 \ \frac{\text{in}}{\text{ft}}}\right)} \\ &= 5000 \text{ ft/min} \end{aligned}$$

This is greater than 2500 ft/min, so use the second equation given.

$$\begin{aligned} L_e &= \frac{\text{v}\sqrt{A}}{10{,}600} \\ &= \frac{\left(5000 \ \frac{\text{ft}}{\text{min}}\right)\sqrt{(36 \text{ in})(24 \text{ in})}}{10{,}600} \\ &= 13.87 \text{ ft} \quad (14 \text{ ft}) \end{aligned}$$

The answer is (D).

72. The absolute pressures before and after compression are

$$p_1 = 0 \ \frac{\text{lbf}}{\text{in}^2} + 14.7 \ \frac{\text{lbf}}{\text{in}^2} = 14.7 \text{ lbf/in}^2$$

$$p_2 = 150 \ \frac{\text{lbf}}{\text{in}^2} + 14.7 \ \frac{\text{lbf}}{\text{in}^2} = 164.7 \text{ lbf/in}^2$$

The absolute temperatures before and after compression are

$$T_1 = 60°\text{F} + 460° = 520°\text{R}$$

$$T_2 = 300°\text{F} + 460° = 760°\text{R}$$

For a polytropic process,

$$\frac{p_2}{p_1} = \left(\frac{T_2}{T_1}\right)^{\frac{n}{n-1}}$$

Substitute values and solve for the polytropic index, n.

$$\frac{164.7\ \frac{\text{lbf}}{\text{in}^2}}{14.7\ \frac{\text{lbf}}{\text{in}^2}} = \left(\frac{760°\text{R}}{520°\text{R}}\right)^{n/(n-1)}$$

$$11.2 = (1.46)^{n/(n-1)}$$

$$\ln 11.2 = \ln(1.46)^{n/(n-1)} = \frac{n}{n-1}\ln 1.46$$

$$\frac{n}{n-1} = 6.38$$

$$n = \frac{6.38}{6.38-1} = 1.186 \quad (1.2)$$

The answer is (C).

73. Use the Mollier diagram in MERM App. "Enthalpy-Entropy (Mollier) Diagram for Steam (customary U.S. units)." For steam at 2000 psia and 400°F, the enthalpy and entropy of the steam at the entrance to the turbine are

$$h_{\text{in}} = 1500\ \text{Btu/lbm}$$
$$s_{\text{in}} = 1.57\ \text{Btu/lbm-°R}$$

For an ideal isentropic efficiency of 100%, the steam would have the same entropy at the exit as at the entrance. From the Mollier diagram, for steam with the same entropy and a pressure of 200 psia, the exit enthalpy is

$$h_{\text{out,ideal}} = 1225\ \text{Btu/lbm}$$

The actual isentropic efficiency is

$$\eta = 0.75 = \frac{h_{\text{in}} - h_{\text{out,actual}}}{h_{\text{in}} - h_{\text{out,ideal}}}$$

The actual exit enthalpy is

$$\begin{aligned} h_{\text{out,actual}} &= h_{\text{in}} - 0.75(h_{\text{in}} - h_{\text{out,ideal}}) \\ &= 1500\ \frac{\text{Btu}}{\text{lbm}} - (0.75)\left(1500\ \frac{\text{Btu}}{\text{lbm}} - 1225\ \frac{\text{Btu}}{\text{lbm}}\right) \\ &= 1294\ \text{Btu/lbm} \quad (1300\ \text{Btu/lbm}) \end{aligned}$$

The answer is (C).

74. A pressure-specific volume diagram for a gas turbine cycle is shown. From A to B is isenthropic compression, from B to C is heating, and from C to D is isentropic expansion through the turbine.

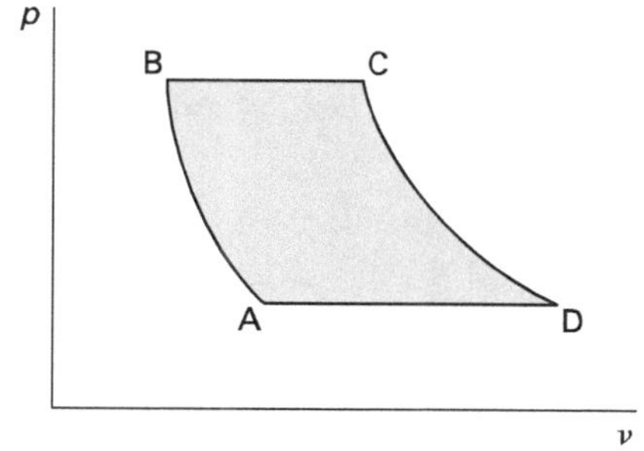

The absolute temperature at state A is $60°\text{F} + 460° = 520°\text{R}$. From MERM App. "Properties of Low-Pressure Air (customary U.S. units)," the reduced pressure, p_r, and specific enthalpy, h, at this temperature are

$$p_{r,\text{A}} = 1.2147$$
$$h_{\text{A}} = 124.27\ \text{Btu/lbm}$$

The reduced pressure at state B is

$$p_{r,\text{B}} = \left(\frac{p_{\text{B}}}{p_{\text{A}}}\right)p_{r,\text{A}} = (5)(1.2147) = 6.074$$

The pressure at state B is

$$p_{\text{B}} = 5p_{\text{A}} = (5)\left(14.7\ \frac{\text{lbf}}{\text{in}^2}\right) = 73.5\ \text{lbf/in}^2$$

Based on MERM App. "Properties of Low-Pressure Air (customary U.S. units)" and a reduced pressure of 6.074, the specific enthalpy at state B is

$$h_{\text{B}} = 196.7\ \text{Btu/lbm}$$

To find the specific enthalpy at state C, calculate the heat of combustion from the heating value of the fuel and the air-to-fuel ratio.

$$q_{\text{combustion}} = \frac{\text{HV}}{\text{AFR}} = \frac{20{,}000\ \frac{\text{Btu}}{\text{lbm}}}{150} = 133.3\ \text{Btu/lbm}$$

The specific enthalpy at state C, the entrance to the turbine, is

$$\begin{aligned} h_C &= h_B + q_{\text{combustion}} \\ &= 196.7\ \frac{\text{Btu}}{\text{lbm}} + 133.3\ \frac{\text{Btu}}{\text{lbm}} \\ &= 330.0\ \text{Btu/lbm} \quad (330\ \text{Btu/lbm}) \end{aligned}$$

The answer is (A).

75. Convert the net work of the cycle to British thermal units per hour.

$$\begin{aligned} W &= (100\ \text{kW})\left(3412\ \frac{\text{Btu}}{\text{kW-hr}}\right) \\ &= 341{,}200\ \text{Btu/hr} \end{aligned}$$

The rate of heat generated from the fuel is

$$\begin{aligned} \dot{Q} = \dot{m}(\text{HV}) &= \left(250\ \frac{\text{lbm}}{\text{hr}}\right)\left(4000\ \frac{\text{Btu}}{\text{lbm}}\right) \\ &= 1{,}000{,}000\ \text{Btu/hr} \end{aligned}$$

The heat available to the cane juice boiler is the heat generated from the fuel minus the heat that is used to generate electricity.

$$\begin{aligned} \dot{Q} - W &= 1{,}000{,}000\ \frac{\text{Btu}}{\text{hr}} - 341{,}200\ \frac{\text{Btu}}{\text{hr}} \\ &= 658{,}800\ \text{Btu/hr} \quad (660{,}000\ \text{Btu/hr}) \end{aligned}$$

The answer is (A).

76. For the gas turbine cycle, out of each unit of energy added to the system, 0.55 unit is exhausted and sent to the heat recovery steam boiler. In the steam cycle, 30% of the 0.55 unit is converted to work, or

$$(0.30)(0.55\ \text{unit}) = 0.165\ \text{unit}$$

The remainder of 0.45 unit is not exhausted by the gas turbine, and so is used for work. The total amount of work for the cycle out of each unit of energy is

$$0.45\ \text{unit} + 0.165\ \text{unit} = 0.615\ \text{unit}$$

The overall efficiency of the combined cycle is 61.5% (62%).

The answer is (C).

77. Use MERM App. "ASHRAE Psychrometric Chart No. 3, High Temperature." Relevant information from the chart is shown.

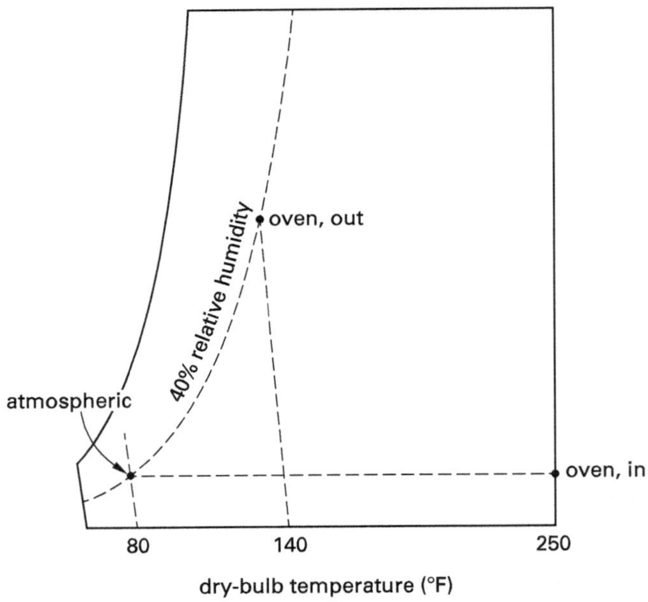

From the psychrometric chart, the humidity ratio of the atmospheric air at 65°F and 40% relative humidity is 0.005 lbm/lbm. Heating the air does not affect this humidity ratio. The specific enthalpy of the entering air at 250°F and a humidity ratio of 0.005 lbm/lbm is

$$h_{\text{in}} = 66\ \text{Btu/lbm}$$

From the chart, for the exiting air at 140°F and 40% relative humidity, the specific enthalpy and humidity ratio are

$$\begin{aligned} h_{\text{out}} &= 94\ \text{Btu/lbm} \\ \omega_{\text{out}} &= 0.053\ \text{lbm/lbm} \end{aligned}$$

While in the dryer, the air increases in enthalpy by

$$\begin{aligned} h_{\text{out}} - h_{\text{in}} &= 94\ \frac{\text{Btu}}{\text{lbm}} - 66\ \frac{\text{Btu}}{\text{lbm}} \\ &= 28\ \text{Btu/lbm} \end{aligned}$$

The air also increases in humidity ratio by

$$\begin{aligned} \omega_{\text{out}} - \omega_{\text{in}} &= 0.053\ \frac{\text{lbm}}{\text{lbm}} - 0.005\ \frac{\text{lbm}}{\text{lbm}} \\ &= 0.048\ \text{lbm/lbm} \end{aligned}$$

The rate of heat gained by the air is

$$\begin{aligned} \dot{Q}_{air} &= \frac{\dot{V}\Delta h}{v_{air}} \\ &= \frac{\left(\left(2000\ \frac{ft^3}{min}\right)\left(60\ \frac{min}{hr}\right)\right)\left(28\ \frac{Btu}{lbm}\right)}{13.33\ \frac{ft^3}{lbm}} \\ &= 252{,}000\ Btu/hr \end{aligned}$$

The rate of moisture gained by the air is

$$\begin{aligned} \dot{W}_{air} &= \frac{\dot{V}\Delta\omega}{v_{air}} \\ &= \frac{\left(\left(2000\ \frac{ft^3}{min}\right)\left(60\ \frac{min}{hr}\right)\right)\left(0.048\ \frac{lbm}{lbm}\right)}{13.33\ \frac{ft^3}{lbm}} \\ &= 432.1\ lbm/hr \end{aligned}$$

The heat lost by the air is picked up by the food product, and the water picked up by the air is lost by the product, so the energy change in the food product is

$$\dot{Q}_{product} = c_{product}\dot{m}_{product}(T_2 - T_1) - \dot{W}_{air}Q_{l,water}$$

$Q_{l,water}$ is the latent heat of water. Solving for the increase in temperature in the food product gives

$$\begin{aligned} T_2 - T_1 &= \frac{\dot{Q}_{product} + \dot{W}_{air}Q_{l,water}}{c_{product}\dot{m}_{product}} \\ &= \frac{252{,}000\ \frac{Btu}{hr} + \left(432.1\ \frac{lbm}{hr}\right)\left(940\ \frac{Btu}{lbm}\right)}{\left(0.75\ \frac{Btu}{lbm\text{-}°F}\right)\left(12{,}000\ \frac{lbm}{hr}\right)} \\ &= 73.1°\quad (73°F) \end{aligned}$$

The answer is (C).

78. Waste heat refers to energy that is generated but not put to practical use. While some waste heat losses from industrial processes are inevitable, losses can be reduced by improving equipment efficiency and by installing waste heat recovery technologies. Waste heat recovery entails capturing and reusing the waste heat from industrial processes or the heat lost from air conditioning systems.

Waste heat can take the forms of both sensible and latent heat. Latent heat is often the greater part, as in combustion gas exhaust.

However, recovering latent waste heat is not as simple as recovering sensible waste heat (high temperature streams). To recover latent waste heat, the transfer material must come in direct contact with the exhausting stream and be able to absorb and evaporate liquids.

Heat recovery wheels, or thermal wheels, come in direct contact with the exhausting stream, and some have the capability to absorb moisture. The evaporation in the heat recovery wheel is a liquid-to-vapor phase change at a constant temperature. A phase change at constant temperature is a latent heat change.

Run-around, heat pipe, and heat pump are indirect methods.

The answer is (C).

79. The equivalent uniform annual cost (EUAC) compares alternatives that have a mix of present, annual, and future costs. The EUAC is the annual cost that is equivalent to the sum of all costs. Present and future costs are converted to equivalent annual costs through the use of cash flow equivalent factors, such as from MERM App. "Cash Flow Equivalent Factors;" these cash flow equivalent factors take interest into account. Then all annual costs and equivalent annual costs are added together to get the EUAC.

The annual electricity cost of the new system is

$$\begin{aligned} C_{elec} &= \left(\frac{110\ hp}{1.341\ \frac{hp}{kW\text{-}hr}}\right)\left(1800\ \frac{hr}{yr}\right)\left(0.12\ \frac{\$}{kW\text{-}hr}\right) \\ &= \$17{,}720/yr \end{aligned}$$

The EUAC for the new system is

$$\begin{aligned} EUAC &= P_{initial}(A/P,\ 10\%,\ 10) + A_{elec} + A_{maint} \\ &\quad - A_{salvage}(A/F,\ 10\%,\ 10) \\ &= (\$80{,}000)(0.1627) + \$17{,}720 + \$2000 \\ &\quad - (\$10{,}000)(0.0627) \\ &= \$32{,}100\quad (\$32{,}000) \end{aligned}$$

The answer is (D).

80. Heat is transferred from the exhaust air to the makeup air. From the psychrometric chart, the specific enthalpy of the makeup air before conditioning is

9.3 Btu/lbm. Revelant information from the chart is shown.

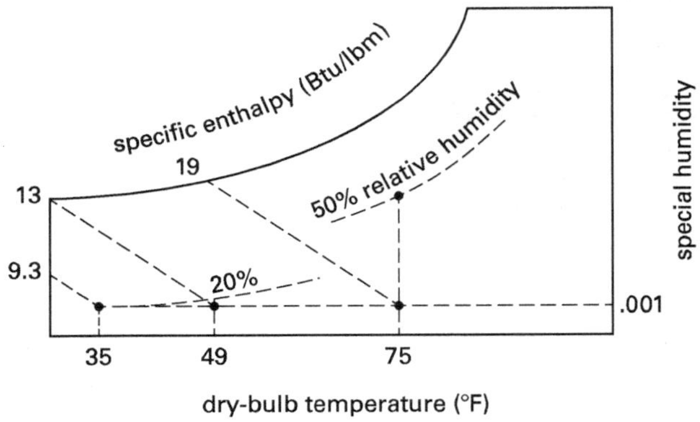

The exhaust air undergoes a latent heat change as it transfers heat to the makeup air. After conditioning, the exhaust air has the same humidity ratio as the makeup air. On the psychrometric chart, find the states of the exhaust air and makeup air before conditioning. Draw a horizontal line to the right from the makeup air, and draw a vertical line down from the exhaust air. The point where these two lines intersect is the state of the exhaust air after conditioning. This point has a specific enthalpy of 19 Btu/lbm. The sensible heat difference between the exhaust air after conditioning and the makeup air before conditioning is

$$19\ \frac{\text{Btu}}{\text{lbm}} - 9.3\ \frac{\text{Btu}}{\text{lbm}} = 9.7\ \text{Btu/lbm}$$

The sensible heat recovered is 40% of this, or

$$(0.4)\left(9.7\ \frac{\text{Btu}}{\text{lbm}}\right) = 3.88\ \text{Btu/lbm} \quad (4\ \text{Btu/lbm})$$

The problem can also be solved more quickly by recognizing that, because the exhaust and makeup airflow rates are equal, a 40% sensible heat transfer results in a 40% dry-bulb temperature rise. The dry-bulb temperature of the makeup air rises from 35°F to

$$(35^\circ\text{F})(1.4) = 49^\circ\text{F}$$

From the psychrometric chart, the conditioned makeup air at 49°F has a specific enthalpy of 13.0 Btu/lbm. The increase in specific enthalpy is

$$13.0\ \frac{\text{Btu}}{\text{lbm}} - 9.3\ \frac{\text{Btu}}{\text{lbm}} = 3.7\ \text{Btu/lbm} \quad (4\ \text{Btu/lbm})$$

The answer is (B).